THE ESSENTIAL GUIDE TO

ROCKY MOUNTAIN MUSHROOMS BY HABITAT

THE ESSENTIAL GUIDE TO

ROCKY MOUNTAIN MUSHROOMS BY HABITAT

Cathy L. Cripps, Vera S. Evenson, and Michael Kuo

University of Illinois Press
Urbana, Chicago, and Springfield

Manufactured in the United States of America
1 2 3 4 5 C P 5 4 3
∞ This book is printed on acid-free paper.
Library of Congress Cataloging-in-Publication Data
Names: Cripps, Cathy Lynn, 1948– | Evenson, Vera Stucky, 1933– | Kuo, Michael, 1963–
Title: The essential guide to Rocky Mountain mushrooms by habitat / Cathy L. Cripps, Vera S. Evenson, and Michael Kuo.
Other titles: Rocky Mountain mushrooms by habitat
Description: Urbana: University of Illinois Press, [2016]
Includes bibliographical references and index.
Identifiers: LCCN 2015027975
ISBN 9780252039966 (cloth : alk. paper)
ISBN 9780252081460 (pbk. : alk. paper)
ISBN 9780252098123 (ebook)
Subjects: LCSH: Mushrooms—Rocky Mountains Region—Identification—Handbooks, manuals, etc.
Classification: LCC QK605.5.R6 C75 2016
DDC 579.60975—dc23
LC record available at http://lccn.loc.gov/2015027975

For Dr. Orson K. Miller Jr.
Mycologist, Mentor, Researcher, Friend

CONTENTS

THE ESSENTIAL GUIDE TO

ROCKY MOUNTAIN MUSHROOMS BY HABITAT

Photo by Cathy L. Cripps

INTRODUCTION

Rocky Mountain Mushrooms in Their Native Habitats

The majestic Rocky Mountains rise to form the backbone of the North American continent, which stretches for 3,000 miles across 34 degrees of latitude from northern New Mexico to Alaska. The great chain is composed of a complex of smaller mountain ranges, each with its own unique geologic structure and history. Along the whole length are: alpine peaks scattered above the tree line, high U-shaped valleys, cirques sculpted by glaciers, an almost continuous band of conifer forests covering steep slopes and valleys from the tree line to the foothills, and at lower elevations, wide expanses of open semi-arid shrublands that merge into wide-open prairie grasslands. Within the forest zone are bands of distinct forest types that can be recognized by their subtle color differences; each band is dominated by one or two tree species and has its own interesting and distinct ecology. The open shrublands

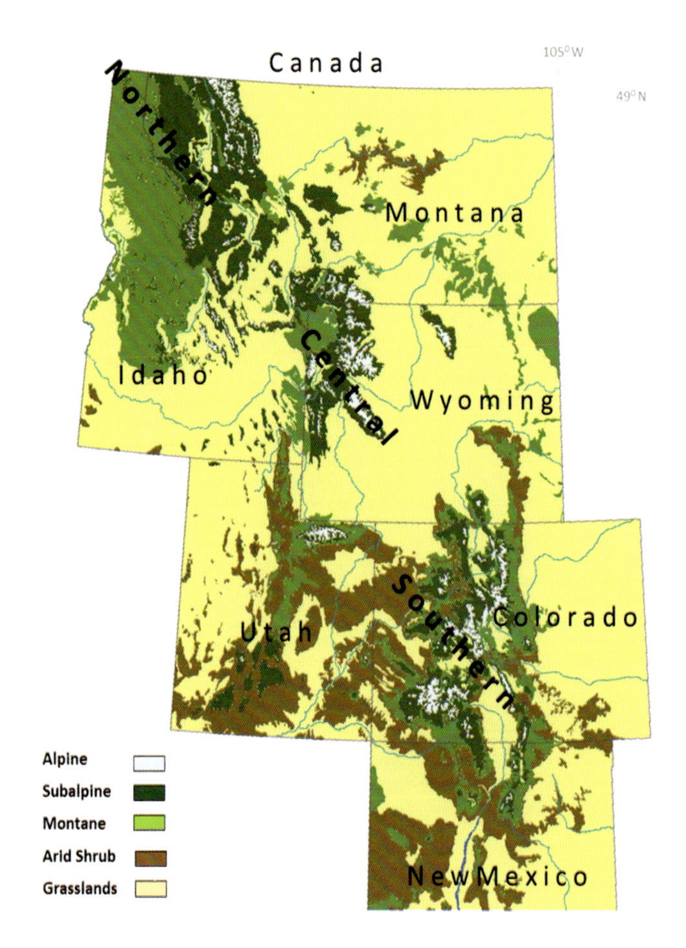

Ecoregion map of the Rocky Mountains, U.S.A.
Map generated by Ed Lubow with USDA FS Ecoregions IV data.

NASA/JPL.

and grasslands that dominate the foothills give these mountains a "western" flavor complete with sagebrush and tumbleweeds.

The climate of the Rocky Mountains is mostly continental; however there is a Pacific Northwest maritime influence in the northern areas and a drier, more desertlike character is apparent at the southern end. Precipitation, as rain or snow, ranges from 40 inches at high elevations down to 10–14 inches in the drier foothills. In general, the foothills become warmer and drier as one goes south toward Mexico. In late summer, monsoon rains provide a reliable source of moisture, which is important for the appearance of mushrooms, especially in the Southern Rockies. The geology of the Rocky Mountains is a hodgepodge of igneous, metamorphic, and sedimentary layers and outcrops, and depending on the location along the chain. Limestone, granitic rocks, and volcanic parent material dominate specific mountain ranges or parts of them. In the south, the tree line reaches to 12,500 feet with numerous mountain peaks above 14,000 feet. In the north, the tree line lowers, and in Alaska it almost reaches sea level where it merges with the northern tree limit that defines the beginning of Arctic habitats.

In general, the Rocky Mountains can be divided into the Northern, Middle, and Southern Rockies. This book is focused primarily on areas within the United States and southern parts of Canada. It can be useful farther north; however, forest types change farther up the mountain chain. In northern Idaho and northwestern Montana, the presence of larch, western hemlock, western red cedar, and grand fir complicates forest types and mushrooms; these tree species are not covered. For our purposes, the Northern Rockies include southern Alberta and northern parts of Montana and Idaho; the Middle Rockies cover most of Wyoming and portions of southern Montana and Idaho; and the Southern Rockies include Colorado, Utah, southern Wyoming, and northern New Mexico.

The Rocky Mountain landscape is a mosaic composed of several dominant vegetation types, which are primarily determined by elevation and moisture regime. We have selected various habitats: grassland prairie; semi-arid shrublands; and forests of ponderosa pine, Douglas fir, aspen, lodgepole pine, spruce-fir, and high pines, plus the alpine for this book. Of course, finer gradations and additional habitats, including mixed forest types, are possible for the Rocky Mountains. For the map, we combined spruce-fir, high pines, lodgepole, and aspen forests for the subalpine, and ponderosa pine and Douglas fir forests for the montane.

Tree species and habitat types gradually change over the length of the Rocky Mountain chain, but some general consistencies can be noted. Much of the character of the alpine with its unique alpine vegetation is consistent north to south, although the dominance of various plants and willow species varies; still, a major portion of Rocky Mountain alpine plants are also known in the Arctic. Much of the Rocky Mountain tree line is dominated by subalpine spruce-fir forests; bristlecone pines in the south and whitebark pines in the north form magnificent and significant forests at the tree line where they are present. At lower elevations, lodgepole pine and aspen are consistent parts of the landscape north to south. However, in montane areas, Douglas fir forests predominate in the U.S. Northern Rockies and ponderosa pine forests are much more extensive in the Southern Rockies, although both tree species occur over the whole distance. Cottonwood riparian is present for most of the length of the Rockies, but the species of *Populus* change from north to south. In the foothills, the character of the semi-arid shrublands changes; pinyon-juniper forests are an important part of the landscape in southern areas but these are absent in the Middle and Northern Rockies. However, sagebrush lands dot the foothills over the whole stretch.

The composition of the mycoflora is influenced by location north to south along the Rocky Mountain chain, as well as by habitat. The species of alpine fungi remain more or less constant along the mountain chain with a few exceptions. Forest fungi in the northern latitudes have a more Pacific Northwest and/or boreal character; while those in the Southern Rockies, especially those in more arid areas, have a

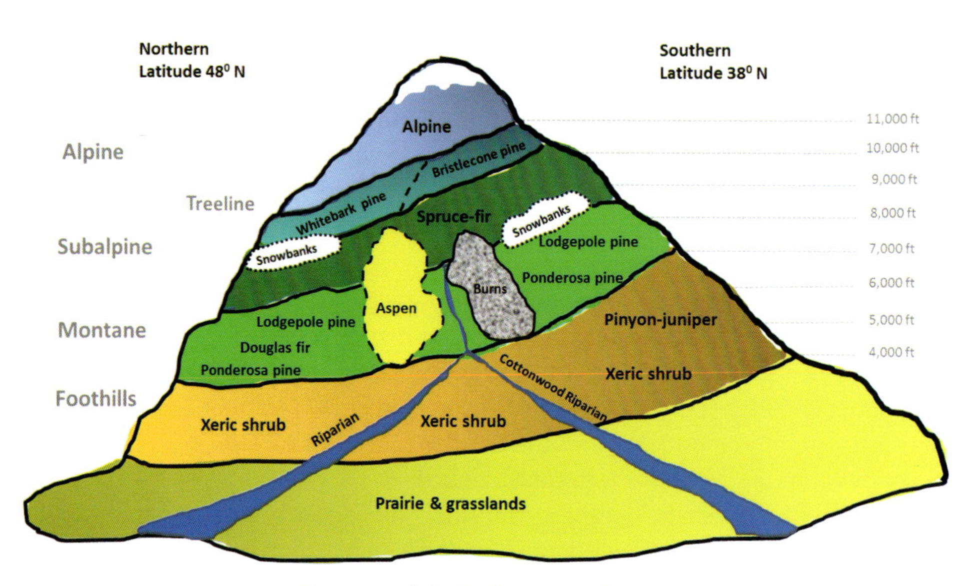

Selected habitats of the Rocky Mountains by elevation. *Diagram by Cathy L. Cripps*

distinctly southwestern flavor like the vegetation. Many unique and unnamed fungal species inhabit this region.

For the high pine forests, we know most about the macrofungi in whitebark pine forests, but we suspect that many of the resident *Suillus* and *Rhizopogon* species might also be found in bristlecone pine forests to the south, since both trees are 5-needle pines that are known mycorrhizal partners with these genera. Interestingly, many of the fungi that occur in lodgepole pine forests, aspen stands, and the cottonwood riparian are relatively constant north to south. In the montane zone, some of the fungi found in ponderosa pine forests in Colorado also appear in Douglas fir stands in Montana. The macrofungi of spruce-fir forests remain relatively constant north to south, although it is worth noting that *Amanita muscaria* has red caps in Colorado (variety *flavivolvata*) and yellow caps in Wyoming and Montana but red caps again in Alaska. Similarly, a main edible bolete in Colorado is *Boletus rubriceps*, but in Montana and Idaho it is *Boletus edulis* (group). Yet the Rocky Mountain chanterelle *Cantharellus roseocanus* appears to span the whole distance from Colorado to Canada. Also, *Cortinarius* species are an important and diverse component of the mycoflora throughout the Rocky Mountain region.

Since each fungus or, extrapolating, each mushroom species is found in its own unique set of circumstances—its native habitat, this information can be useful for finding and recognizing particular species. The Rocky Mountains lend themselves to this approach since basically habitats are distinct and divide out nicely by elevation and dominant vegetation type. While not a perfect system, this is often how we think about our mushrooms in the Rocky Mountains. It is why we go to the cottonwoods in spring for yellow morels, or to burns for black morels in June, or to conifer woods for boletes and tooth fungi in autumn. In contrast, the mixed deciduous forests of the East and mixed conifer forests of the West Coast appear to be an indecipherable cacophony of hosts and fungi to Rocky Mountain folks who are more familiar with roaming around in simpler habitats composed of one or two dominant tree species.

This book is a mycological tour that starts in the sweeping prairie grasslands at the base of the Rockies and moves through the foothills of semi-arid shrublands and cottonwood riparian, up through conifer and aspen forests at higher elevations, and all the way to the airy alpine with its miniature forests of dwarf willow and nano-mushrooms. We also include habitats that support the unique sets of fungi found on burned ground after wildfire and those that fruit next to melting snowbanks in spring. This book is a mycological travel guide that gives a tantalizing sample of the fungal inhabitants found at each stop on the tour. We hope you enjoy the journey.

Ecological Roles of Fungi

Fungi provide an array of ecosystem services that include, but are not limited to, decomposing and recycling a variety of plant materials; sequestering carbon in soil; providing nutrients to trees and plants through their roots; holding soil in place with their mycelium; and providing food for invertebrates and mammals as part of the food web. While fungi have a diversity of ecological roles in nature, these could be boiled down to: decomposers (saprobes), mycorrhizal mutualists, and pathogens. We have tried to mention the ecological role of each of the mushroom species in our book to make the point that, as in a well-organized society, each fungus has its own role to play.

Saprobic fungi are nature's recyclers, and each year they break down and recycle millions of tons of plant matter along with other material. These fungi are usually specialists to some degree and each has dietary preferences; some prefer conifer

The actual fungus body is composed of fine threads called mycelium, shown here decomposing wood and serving to hold burned soil together. *Photos by Cathy L. Cripps, Don Bachman*

Brown cubical rot in a conifer log and the brown rot fungus *Gloeophyllum sepiarium* fruiting. *Photos by Cathy L. Cripps, Vera S. Evenson*

wood, some hardwoods; some fungi decompose only cones—and of certain trees, some decompose twigs, needles, or leaves, and then there are those that prefer to dine on dung. Each species has its own special niche and this is where we learn to search for them.

Brown rot fungi decay mostly conifer wood, decomposing the cellulose and forming residues that are in the form of tiny brown cubes. We see their work in the crumbly remains of fallen trees. Other fungi turn mostly hardwoods such as aspen into a stringy white residue; these are the **white rot** fungi and they preferentially decompose lignin, the tough stuff of wood.

Dung fungi take advantage of the quantities of plant matter left over in cow, horse, elk, deer, moose, goat, and sheep droppings and make this fertile habitat their home. Saprobic meadow mushrooms (*Agaricus* spp.) decompose dead grass and soil organics and produce fairy rings as their mycelium radiates outward in the soil.

Mycorrhizal fungi attach themselves to the roots of plants where they live on the sugars provided by their host, and the fungi return the favor by using their bodies

A dung-loving *Conocybe* sp. on bison dung.
Photo by Cathy L. Cripps

Fairy ring (*Mycenastrum corium*) in Wyoming grasslands.
Photo by Cathy L. Cripps

(mycelium) as conduits to move nutrients (nitrogen, phosphorus) from the soil into plants. Because the fungus and the plant both benefit from this association, it is called a **mutualism**. In nature over 85 percent of all plants are mycorrhizal, but only some fungi, called **ectomycorrhizal fungi** are capable of producing fruiting bodies. The **ectomycorrhizae** they form with their mycelium can be observed on fine root tips of the plant host with a hand lens.

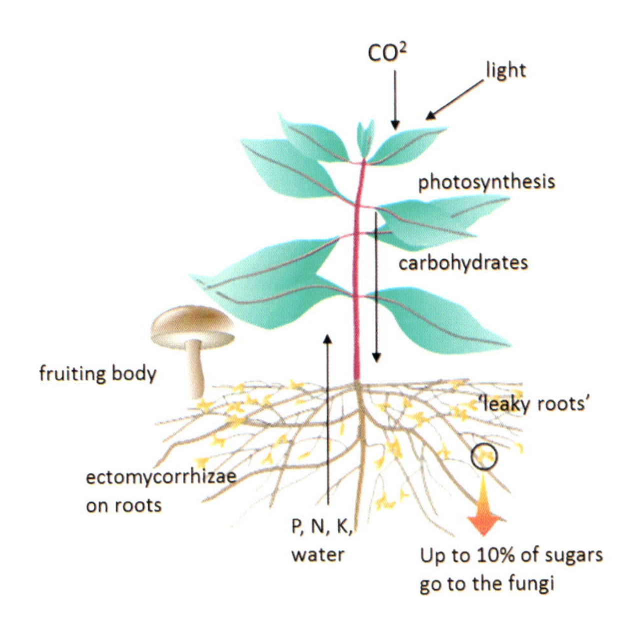

How mycorrhizal fungi work. *By Cathy L. Cripps*

Mycorrhizae on the roots of Bearberry (*Arctostaphylos*).
Photo by Cathy L. Cripps

Foliose lichen on aspen bark and crustose lichen on rock. *Photos by Cathy L. Cripps, Loraine Yeatts*

Some ectomycorrhizal fungi attach themselves to Douglas fir roots, while others prefer aspen or pine roots, as there is often (but not always) some host-fungus specificity involved. Each individual tree has a variety of mycorrhizal fungi on its roots at any one time, and these change over the life of the tree from **early colonizers**, which prefer seedlings, to **late colonizers**, which favor mature trees. In some cases, we can use this information to find the fungi we desire by visiting the appropriate age forest.

Rust fungus *Melampsora epitea* on dwarf willow *Salix reticulata*. *Photo by Cathy L. Cripps*

Shoestring root-rot fungus *Armillaria solidipes*, a Rocky Mountain species. *Photo by Vera S. Evenson*

Another type of mutualism is the **lichen lifestyle**. These "dual organisms" are composed mostly of fungal hyphae with a layer of algae (or cyanobacteria) sandwiched into the top layer. Algae produce sugars that feed both themselves and the fungi through photosynthesis. In this mutualism, fungi provide living space for algae that would find it hard to go it alone. The upshot is that lichens are able to make their living on air (CO_2), rainwater, and dust (nutrients). Lichens can be observed attached to tree bark, branches, rocks, and soil. Their colorful appearance is due to the algae and to the pigments produced by the fungi, which protect the organism from harmful UV light. After all, they sometimes sit exposed for hundreds of years.

Fungal pathogens can have positive as well as negative effects on plants and forests. True, they attack living organisms and find their way around the defense systems of plants and animals to invade living cells. But when a group of trees is killed by these fungi, **forest gaps** appear and succession is reset so that a new diversity of life can continue in these openings. But the balance between host and pathogen has been upset time and time again with the introduction of exotic pathogenic fungi from faraway lands for which hosts have no resistance; invasions can become widespread. This is true for the white pine blister rust introduced from Europe, which currently threatens western 5-needle pine forests in western North America.

How to Use This Book

This book is for those who would like to know more about the mushrooms they observe in fields and forests. It is intended to make the names of certain mushrooms found in particular habitats easily accessible to naturalists, forest managers, casual observers, and mushroom aficionados alike. We also elucidate a few of the fascinating ways plants, animals, and fungi interact in their respective native habitats.

This book is also for those already initiated in mushroom collection and identification; it contains some special species and provides an alternative way of learning—*by observation of mushrooms in their specific native habitats*. To streamline the identification process, this approach highlights the mushrooms expected in a particular habitat. We include representative and interesting species for each section; however, it is important to know that not all mushroom species could be included for each habitat and that more detailed field guides may be needed to confirm a tentative identification.

The best way to approach this book is to *learn to recognize the habitat* in which you are collecting. Knowing the dominant forest trees or vegetation type and approximate elevation is essential. The main tree species are described within each section; luckily, the number of tree species is limited in the Rocky Mountains. After identifying your habitat type, you can simply flip through the pages of the selected section to see whether your mushroom or fungus is included. In some cases, a mushroom might be found in several habitats (especially true for conifer forest types) and you can expand your search to other sections using the Quick Guide inside the front cover.

Each habitat is *color-coded* and species are listed in alphabetical order within each of these groups: gilled mushrooms (light, then dark-spored), boletes, nongilled fungi, gasteroid fungi, and Ascomycota. At the back of the book, we provide a key with fungi grouped by their morphology (form) to make it easier to see which species in each group are included in the book, i.e., all polypores are grouped together as are all tooth fungi. Some mushrooms are found in several habitats, and the Quick Guide is helpful for finding where an individual species is placed in the book. The section "Further Reading and References by Habitat" provides detailed sources.

The All-Important Spore Print Color

Knowing the spore color of a *gilled mushroom* is particularly helpful in its identification. Spore color is not always the same as gill color, and a mistake here can rapidly throw off progress toward the correct identification. There are several ways to determine the spore color of a mushroom:

1. **Do a spore print at home.** Simply cut off the stalk of a fresh mushroom, place the cap gill-side down on a piece of paper and cover it with a bowl or glass to fit the cap size. If all goes well, in a few hours or overnight, a "print" of the spores will appear on the paper. The biology is that in the humid chamber, spores are shot from small cells (basidia) on the gills, which land on the paper. If you blacken part of the white paper with a felt pen first, it can help in viewing white spores later on.
2. **Observe in the field.** Notice whether the spore color can be observed on mushroom caps that are growing overlapped, with spores falling on the lower cap. Spore color can sometimes also be discerned on surrounding soil or vegetation. In that case, include a piece of the spore-covered substrate in your collection.
3. **Observe the mushroom.** Spores tend to stick to rings or stalks, and spore color can sometimes be determined by a close look at the mushroom.
4. **Start a spore print in the field.** Place a cap (with stalk removed) on a small piece of paper and wrap it in aluminum foil or place it in a small plastic box, such as a tackle box, so it can print before arriving home. It is important that the cap remain face down during this process.
5. **Use a microscope.** While spore color looks different when viewed with a microscope, with experience this can be useful.

The next step is to be able to recognize the spore color on the paper. For our purposes the categories are: light (white, cream, yellow); pink (usually salmon pink); or dark (yellow brown, dark brown, medium brown, rusty brown, black brown, purple black, or black). If you know the spore color, it will help you hone in on particular species, ignore others, or confirm a genus or group.

Light-colored spore print from a *Russula* in the field. Examples of a light spore print (white, cream, or yellow) on felt-pen darkened areas, a strong pink spore print, and a dark spore print (yellow brown, brown, purple brown, or black).
Photos by Vera S. Evenson, Cathy L. Cripps

FOOTHILLS

Photo by Loraine Yeatts

THE AMERICAN PRAIRIE

The American Prairie covers many thousands of square miles of grasslands that sweep gently eastward from a series of spectacular North American mountain ranges collectively known as the Rocky Mountains. Ancient volcanic activity, regional uplifts, glaciations, erosion, and climate change shaped our world-renowned mountain formation. This immense crest of rocks is a continental divide in a most literal sense; the extreme heights of the massive Rocky Mountain ranges and high peaks running roughly north and south divide a large part of our North American continent.

The climate and geography of our prairies and plains formed on the east and the west sides of the Rockies are a direct result of this uplift. Stretching for more than 3,000 miles from northern British Columbia in western Canada down to the Rio Grande in New Mexico, the extensive uplift of rock has created a rain shadow by robbing the western winds of their moisture. The result is a beautiful sea of grassy prairie lands known for their characteristic lack of forests. Instead of forest green,

Subtle colors of the prairie. *Photo by Ken Evenson*

Today wild bison are restricted to a few grasslands, primarily Yellowstone National Park. *Photo by Loraine Yeatts*

prairies show us a mosaic of soft browns and gray greens, punctuated with bright spots of colorful native wildflowers.

Sometimes the rolling sea of grasses, herbaceous forbs, and shrubs is broken by a ravine or wandering waterway. All the plants there as well as the birds and other animals have evolved by adapting to the semi-arid soils and harsh extremes of weather.

Amid waving grasses and prairie wildflowers, this landscape speaks to us of our human heritage. For thousands of years, native peoples traversed the immense grasslands in order to follow the great herds of bison that found their perfect home in the vastness. Later, our pioneer ancestors came through these rolling grasslands in search of a new life, eventually settling in many areas, usually near rivers and other water sources. The name for these vast grasslands or meadows was adopted by these hardy settlers from the early French explorers' use of their word for meadow or grassland, which was *prairie*.

Although a large percentage of the millions of acres of North American prairie have since been plowed for croplands or grazed by domestic animals for the past century or more, some priceless remnants remain in the prairies of Montana near Malta, the eastern Washington and Idaho grasslands known as Palouse country, and prairie land in the eastern parts of Utah, Wyoming, Colorado, and New Mexico.

A rich mosaic of native grasses grows in these semi-arid habitats. Short and mixed grasses, the hallmarks of the western prairie, have provided nutritious fodder for its native animals and birds throughout history. Pronghorns, celebrated for their long-distance running skills, evolved in these great expanses of grasslands by outwitting the now-extinct native cheetahs to survive to this day in large herds. Until recently, there were more pronghorn individuals in the state of Wyoming than humans. Their range still includes prairie lands throughout the Rocky Mountain region.

Prairie evening primrose and purple prairie aster, near Pawnee National Grasslands, which remain as a living short grass treasure. *Photo by Loraine Yeatts*

Pronghorn antelope in the grasslands. *Yellowstone National Park Photo*

The dominant two grasses in the short-grass prairie are buffalo grass (*Bouteloua dactyloides*) and blue grama (*Bouteloua gracilis*). These and other native grasses have adapted to the extremes of variable moisture and harsh weather conditions by developing exceptionally deep roots that penetrate the soil as far as 10 feet down. Prairie grasses thrive in semi-arid conditions by a water-use strategy that involves rapid intake of water when it is available and then, when it is scarce, they become dormant.

Buffalo grass and blue grama are dominant prairie grasses. *Photos by Loraine Yeatts*

Bush morning glory (*Ipomoea leptophylla*).
Photo by Loraine Yeatts

Adding a delightful contrasting color to the prairie, indigenous herbaceous species have also adapted to the low moisture and intense sun in this semi-arid habitat. The bush morning glory survives here by developing turnip-shaped roots that grow deep in the soil, enabling it to overcome droughts, overgrazing, and prairie fires.

Also penetrating the deep prairie soils are the underground tunnels made by villages of social rodents known as prairie dogs. The journals written by Lewis and Clark during their famed expedition to the West in 1804 made note of their first discovery of a prairie dog village in Nebraska. A live specimen was captured and sent back to President Jefferson in the White House, much to his delight.

Commonly referred to as a "keystone" species in many western American prairie habitats, the black-tailed prairie dog and its huge colonies have had a great impact historically on the prairie grassland environment. The foraging and burrowing action of many millions of these communal rodents has enhanced the diversity of the grasses and herbaceous plants there. Animals native to these habitats, both vertebrates and invertebrates, have benefited from the constant disturbance and aeration of the deep prairie soil caused by these gregarious natives.

Burrowing owls inhabit open grasslands, preying on a variety of insects and small animals such as mice and lizards, and often nest in abandoned prairie dog burrows. Spotting a charming group of these uncommon owls is a rare treat, which is becoming rarer as their habitat and prey gradually disappear.

The mountain plover (*Charadrius montanus*) is a characteristically short-grass prairie bird, living exclusively in open grasslands where it nests in a scrape on bare ground, historically near heavily grazed bison habitats or prairie dog farms where grass is short or trampled. Both of these native birds' populations have been greatly diminished during the past century because "keystone" species such as prairie dogs and other interdependent native species have been greatly reduced in numbers with overgrazing by domestic animals, plowing under native soils, and other human activities.

Black-tailed prairie dog (*Cynomys ludovicianus*) on alert at its mound.
Photo by Rick Brune

An observant family of burrowing owls (*Athene cunicularia*). *Photo by Richard Holmes*

The mountain plover is a misleading name since this is a bird of the prairies, not the mountains. *Photo by Stephen R. Jones*

Too moist to be deserts, too dry for forests, the prairie grasslands support a sparse and sporadic mushroom flora. Most of those fungi are saprobes that break down and recycle the dead grasses and forbs in the vast prairie lands, thus helping to build soils and recycling nutrients for the use of native plants that eventually provide fodder for the animals that live there. In turn, the dung from those animals is eventually recycled by specialized coprophilous (dung-loving) fungi, thus continuing the constant interchange of nutrients and organic materials in this amazing habitat. Native fungi play an absolutely essential role in the harmony and balance of the prairie ecosystems.

Dung fungus (*Deconica coprophila*) decomposing bison dung. *Photo by Ed Barge*

Fairy rings (*Agaricus bernardii*) in open grassland. *Photo by Vera S. Evenson*

Storm on the prairie brings up mushrooms. *Photo by Don Bachman*

Grassland prairies present flat unobstructed areas for mushrooms to show up as fairy rings and arcs that are sometimes many meters across. Folklore in many cultures suggests that the mystery of mushrooms growing in circles and killing the grass inside the circles can be explained by fairies, leprechauns, or even dragons dancing in magic circles at night. A more scientific explanation is that the original mycelium develops from a spore and grows outward in all directions, sending up fruiting bodies at the periphery in the form of huge circles. This process continues, however slowly, sometimes for as long as a century if the soil is not disturbed. Often the fungal mycelial growth inside the circle is so dense or so devoid of nutrients that the grass inside the circle dies back, forming a change in color or aspect of the grass that is hard to explain. A variety of fungi form rings including *Agaricus* species, *Marasmius oreades*, and a whole host of large puffballs.

Photo by Cathy L. Cripps

Amanita prairiicola Peck

DESCRIPTION: CAP 6–20 cm across, convex to flat; white or cream, with flat innate warts that make for a bumpy appearance; sticky, with soil adhering; margin turned in at first, hung with patches of white tissue, later splitting. **GILLS** free, crowded, broad; pale cream to dingy gold; edges floccose, eventually eroded. **STALK** 15–20(25) × 2–4 cm, enlarging to a swollen onion-bulb base; white to cream; sticky, with ragged zones of tissue below a fragile pendant ring; deeply buried in soil; bits of tissue (volva) on the base. **FLESH** white, solid in stem; odor strong and unpleasant. **SPORE PRINT** white.

ECOLOGY: In open cultivated fields and sagebrush areas, typically without a mycorrhizal host; fruiting in early to late spring. Known from Oregon, Montana, Idaho, Wyoming, and Colorado, but originally described from Kansas. Here, we report it from an open cultivated field with no potential for a mycorrhizal host, unusual for an *Amanita*.

OBSERVATIONS: *prairiicola* for its habitat. This *Amanita* is unusual for its lack of a need for a mycorrhizal host plant and it could be mistaken for a *Macrolepiota*. It appears to be the same as *A. malheurensis* Trueblood, O. K. Miller Jr., and Jenkins (Miller et al. 1990). Ellen Trueblood was a well-known amateur mycologist from Idaho who collected with A. H. Smith and O. K. Miller. Not edible.

Photo by Vera S. Evenson

Clitocybe praemagna (Murrill) H. E. Bigelow and A. H. Smith

DESCRIPTION: CAP large, 10–30 cm across, convex; dull pale brownish; surface fairly smooth and dry, often covered with dirt from eruption through the soil; margins with tiny cracks with age. **GILLS** white to cream, close, attached-adnate. **STALK** short and bulky, 2–6.5 cm long × 2–5 cm across, equal to slight bulb at base; surface lightly scurfy; white with pale cinnamon-colored area; without veil remnants. **FLESH** firm, white; odor and taste mild. **SPORE PRINT** pinkish.

ECOLOGY: Erupting through the soil; a saprobe in grasslands and open meadows, often in arcs or fairy rings, late spring into summer after thunderstorms.

OBSERVATIONS: *praemagna* for the large size of the fruiting bodies. This mushroom is distinctive because of its manner of growth, bulky size, and tendency to grow in arcs or clusters. It is not common but has been reported from Wyoming, Idaho, Colorado, and Montana, as well as the Pacific Northwest and Canada, always in open grasslands or former prairies. Smith et al. (1979) reports it as edible and from western grasslands including sagebrush areas. Great care is needed to confirm the identification. Called *Lepista praemagna* by mycologist Rolf Singer.

Photo by Ed Barge

Marasmius oreades (Bolton) Fries

DESCRIPTION: CAP 2–5 cm broad, distinctively campanulate, becoming more or less plane, often with rounded umbo; surface smooth, dry; margins slightly wrinkled with age; colors dull orange-cinnamon to salmon to pale ocher-brown, fading upon exposure to light. **GILLS** distant, broad, adnexed to nearly free; pale cream. **STALK** 4–8 cm long × 4–8 mm wide, equal; pale ochraceous above with reddish brown colors at base; tiny hairs near base; texture distinctively tough and not breaking easily. **FLESH** rather firm; pale ochraceous to buff; odor mildly fragrant, often like cyanide; taste mild. **SPORE PRINT** white.

ECOLOGY: Known as the fairy ring mushroom, *Marasmius oreades* is a common saprobe in the soil of grassy areas of such diverse habitats as backyards, native prairies, golf courses, or city parks, often forming fairy rings. *M. oreades* can be found there throughout the growing season, most abundantly in warm weather after rains.

OBSERVATIONS: *oreades* for the Greek expression for mountain fairies. Members of the genus *Marasmius* are known for the ability of the fruiting bodies to revive after drying and even produce spores again. Edible, but caution is advised as toxic species can grow among them or form their own fairy rings.

Photo by Vera S. Evenson

Agaricus bernardii Quélet

DESCRIPTION: CAP large, 5–12 cm across, hemispheric when young, becoming convex soon, flattened on top and often depressed; margins exceeding the gills, remaining incurved; surface white to dingy buff; dry, at first smooth, often developing brownish flattened scales with age; veil remnants at margins whitish, cottony. **GILLS** grayish pink in youth, becoming chocolate-colored; close, free, and narrow. **STALK** 4–9 cm long × 1.5–3 cm, commonly narrowing at base; veil thin, white, membranous, sheathing the lower stalk, soon tearing as cap expands, leaving a thin, torn edge resembling white tissue paper on stalk midsection and on cap margin. **FLESH** firm, white, immediately turning reddish to brown-orange when cut; odor pungent; taste rather strong. **SPORES** in mass dark brown.

ECOLOGY: Growing as a saprobe in soil in groups or fairy rings; open areas, prairies, parks, and pastures; fruiting in midsummer to early fall after rains.

OBSERVATIONS: Distinguished in the field by its stout stature, rufescent flesh, thin cottony veil, and gregarious growth habit, this species often fruits in sandy or saline soil. *Agaricus bitorquis* is similar with its short, stout stature and strongly inrolled cap margin, but its flesh is unchanging at injury and its odor is mild. It fruits along roadsides, sometimes in hard-packed soils.

Photo by Cathy L. Cripps

Agaricus campestris Linnaeus

DESCRIPTION: CAP 2.5–8 cm across; hemispheric to convex, then almost flat; smooth or with faint scales, dry; white or a bit tan; margin turned in at first. **GILLS** free, crowded; bright pink when young, then grayish pink, finally dark chocolate brown. **STALK** 2–5 × 1–2 cm, equal and usually tapering to a point; smooth, silky, dry; bright white; with a ring that is only a bit of ragged tissue and not skirtlike. **FLESH** white, turning slightly brownish when cut open; odor fungoid. **SPORE PRINT** dark chocolate brown.

ECOLOGY: Scattered or in rings in grasslands and meadows, from the prairie to the alpine; widespread, commonly fruiting in spring, particularly in June but also in the fall after rain; a decomposer of organic matter in soil, including that from dead grass.

OBSERVATIONS: *campestris* for fields or plains. The well-known "meadow mushroom" or "pink bottom" can be recognized by the pointed stem base, flesh that barely turns brown when cut, the very thin ring, and a weak odor. The much larger *Agaricus arvensis* has a skirtlike ring, stains a bit yellow and has a faint almond odor. There are many other *Agaricus* species in open grasslands, especially in Colorado, and many are not well-known or named, so their edibility is also unknown. This species is an excellent edible, but the identification should be carefully confirmed.

Photo by Ed Barge

Polyporus cryptopus Ellis and Bartholomew

DESCRIPTION: CAP small, 0.5–1.5 cm across, shallow convex; pale brown, buff, cream; smooth, kidskin dry; margin turned down. **PORES** go slightly down the stalk, radially arranged, somewhat hexagonal; small, 2–3 per mm; cream; mouth of pores appears ragged or with small hairs (ciliate). **STALK** 0.8–1.2 × 0.1–0.2 cm, equal, rough ridged; dingy white at top, black for most of the length. **FLESH** tough, woody in both cap and stalk; odor not distinctive. **SPORES** white.

ECOLOGY: Appearing terrestrial or attached to "woody" rhizomes of grass or possibly sagebrush belowground; a decomposer of woody material in prairies, known from the American Prairie Reserve in Montana, Colorado, and the Great Plains. On a global scale as *Polyporus rhizophilus*, it is reported on "steppe grasses" in Europe and Asia.

OBSERVATIONS: This "prairie polypore" is rare or overlooked in prairie habitats. It is similar to *P. badius* (also known as *Royoporus badius*) with its black stem and the pores are a bit hexagonal like *P. arcularius*, but the habitat is different.

Photo by Vera S. Evenson

Calvatia bovista (Linnaeus) Persoon

DESCRIPTION: SPORE CASE large, 10–25 cm across, 10–20 cm high; at first rounded on top, then flattening somewhat; with distinctive narrow, tapered base; surface white, outer skin soon cracking into brownish wartlike flakes; as those flakes fall away, a pale brown parchmentlike inner skin is revealed, which erodes with age to free the spore mass for dispersal. **STERILE BASE** extensive, occupying nearly one-half of the puffball, chambered; outer surface and base eventually smooth; white at first then brownish; enduring in the habitat as a persistent, vaselike hull long after the spores are dispersed. **GLEBA** (interior) solid at first, white, then aging to yellow-brown and finally to olive-brown as the spores mature; odor and taste mild. **SPORES** yellow-brown.

ECOLOGY: A saprobe, this large puffball is found scattered to occasionally numerous in open fields and prairies. Widely distributed, reported from Alberta to New Mexico in our Rocky Mountain region. Fruiting in summer to early fall.

OBSERVATIONS: Another large puffball found in our region's grasslands and steppe habitats is *Calvatia booniana* (Pg 41). Its larger oval to flattened shapes and lack of a sterile base differentiate it from *Calvatia bovista*.

Photo by Vera S. Evenson

Calvatia cyathiformis (Bosc) Morgan

DESCRIPTION: SPORE CASE 5–19 cm across and 8–15 cm high, often pear-shaped with a tapered sterile base; outer surface at first whitish tan becoming brown, soon cracking irregularly and flaking off as it ages. **STERILE BASE** chambered, prominent, occupying most of the narrow lower part of the fruiting body, often persisting as vase-shaped remnants when the spores have been dispersed. **GLEBA** (interior) at first whitish, becoming yellow grayish, finally colored purple-brown as spores mature.

ECOLOGY: A characteristic occupant of prairie grasslands, fields, and desert communities, at times extending up into the semi-arid foothills of the Rockies; summer to fall. A saprobe, this easily recognized, purple-spored, fairly large puffball has been reported to form huge fairy rings in the prairies of Colorado.

OBSERVATIONS: *cyathiformis* for the urn-shaped remnant of the sterile base that often persists for many months after the spores have blown to the winds. *Calvatia bovista* (Pg 27) is a similar prairie inhabitant that also eventually produces empty sterile bases, but its spore mass is distinctively dark olive-brown.

Photo by Vera S. Evenson

Disciseda bovista (Klotzsch) Hennings

DESCRIPTION: SPORE CASE 1.5–2 cm across, shaped like spherical acorns, mature specimens become flattened with a rounded, ragged more or less central pore; surface of the spore case roughened by mycelium, becoming smooth, papery, and pale gray beige in color. **BASE** of the spore case is a conglomeration of sandy soil and mycelium that remains attached as a sand case. **STALK** absent. **FLESH** firm at first, whitish then maturing to a powdery brownish maroon mass of spores; odor and taste mild.

ECOLOGY: Terrestrial, scattered in sandy or granitic soils, found in the high rolling plains and open grasslands of many semi-arid parts of the Rocky Mountains, often near buffalo grass; late summer persisting into the next spring.

OBSERVATIONS: These interesting tough little fungi develop under the shallow prairie soil surface, with mycelium covering the developing spore case. As it matures, the entire spore case becomes detached from the soil by the movement of the wind and then is exposed to the harsh prairie environment. The spore case gradually rolls over so its former soil-covered top becomes the base—a heavy sand case that helps hold the spore case upright on the surface of the prairie soil like a harbor buoy. A pore develops in the exposed papery upper "skin" so the spores can escape by the slight wobbling movement brought about by the ever-present breezes among the prairie grasses.

Photo by Vera S. Evenson

SEMI-ARID SHRUBLANDS

The semi-arid shrub/woodlands at the foot of the Rocky Mountains are a sweep of open landscapes defined by the scarcity of a life-giving resource—water. All of the plants, animals, and fungi of these dry places have learned to live with low rain and snowfall. This is a landscape of buttes, plateaus, and cliffs; of canyons, ravines, gullies, and arroyos; of dry alkali flats cut through by large rivers or ephemeral streams punctuated by precious seeps and springs that provide oasislike respite. Sculptured sedimentary sandstones of red or yellow add color that contrasts with the sparse pale green vegetation.

Sunrises and sunsets come and go with the gold and orange-striped palettes characteristic of our semi-arid western lands. The well-spaced almost desertlike vegetation is punctuated by dense thickets and patches of small trees, low evergreens, and sagebrush. Here and there, especially in the Southern Rockies, signs of ancient people are evidenced by the ruins of their rock houses and petroglyphs on rock faces.

The semi-arid shrublands cover a significant portion of the Rocky Mountain foothills where rainfall is limited to 10–20 inches per year. This habitat type is found on the Colorado Plateau, which includes parts of Colorado, S. Idaho, and Utah, and it is plentiful in New Mexico, Wyoming, and east of the continental divide in Montana.

A slight rise in elevation from many western shrublands will put you into the orchardlike pinyon-juniper woodlands, nicknamed "P-J" by the locals. Often called pygmy forests, the vegetation there is indeed short in stature; but the tough drought-resistant trees and shrubs are powerful in their long-lived presence in this habitat. Rarely growing over 30 feet tall, both pinyon pines, *Pinus edulis* and *P. monophylla*, and one or more of the species of *Juniperus* are indicator tree species in this loose-soiled, often rocky landscape. Life-giving by-products of these two main types of trees are juniper berries and pine nuts, food sources for almost all animals and many birds that live there—or that eat those animals that live there. Juniper berries, which are actually much modified cones, are a food source for many wildlife species, including thrushes, grouse, chipmunks, foxes, and deer. The aromatic wood is often used for cedar chests as well as fence posts, lumber, and fuel.

Pinyon trees with their bushy stature grow on dry rocky foothills and mesas and are a very important source of edible seeds, food for birds and animals such as woodrats, deer, wild turkeys, and even bear. Often in flocks of up to 50 or more, raucous pinyon jays serve themselves when they harvest seeds from the ripe pinyon cones, and they serve the pinyon trees by "planting" the seeds during their nomadic stockpiling efforts. It has been said that "tree feeds bird, bird plants tree."

The word *pinyon* comes from the Spanish word *piñon* meaning, *nut*. Throughout their long history, indigenous peoples have harvested the nutritious pinyon nuts for food and in more recent times for sale in local markets.

Native grasses and shrubland plants are interspersed among the pines and junipers and, in many areas of the Rocky Mountains, Big Sagebrush is the dominant plant. Other areas of the semi-arid shrublands are punctuated with dense thickets

Open pinyon-juniper woodlands. *Photo by Lee Gillman*

Pinyon jays eat the seeds of pinyon pines and also plant the seeds. *Photo by Joseph Mahoney*

Sagebrush (*Artemesia* spp.) country in Wyoming.
Photo by Cathy L. Cripps

of gambel oak (Southern Rockies), mountain mahogany (*Cercocarpus* spp.), both ectomycorrhizal trees, and other shrubs such as greasewood, saltbush, and creosote bush.

A day wandering these lands can be full of surprises if one is attuned to the nuances of life on a dry landscape. Morning brings precious dewdrops to leaves that quench the thirst of small animals such as insects. The colorful collared lizard (*Crotaphytus collaris*) moves out onto rocks in the early light to warm its cold blood by doing push-ups so it can scurry around. Rattlesnakes can be observed warming themselves on sun-drenched rocks later in the day.

Bighorn sheep with muscular bodies and rams with massive horns awake from their beds and climb nimbly among the rocky outcrops or sit and chew their cuds. Black-tailed jackrabbits zigzag across the land to find a meal of dried plants without getting caught by predators. Like the sheep, they are one of the few mammals that are not nocturnal and in the heat of the day they must find shade, although their large ears function to dissipate heat. Most mammals in the semi-arid shrublands are nocturnal and hide or burrow during the day. But in the cool of the evening, coyotes and bobcats begin their prowl for small rodents who must reveal themselves to find food.

Scavengers such as ravens and vultures perched high on cliffs help complete the food cycle; they await the opportunity to pick clean carcasses left by predators. Nothing goes to waste here and only bleached white bones will eventually mark the

Thickets of gambel oak (*Quercus gambelii*) in the Southern Rockies and oak leaf close up. *Photos by Michael Kuo*

The brilliantly colored Collared Lizard warming itself on a rock.
Photo by Loraine Yeatts

Bighorn sheep (*Ovis canadensis*) resting in sagebrush country.
Photo by Cathy L. Cripps

spot. While vultures are silent fliers, the ravens announce their presence with noisy caws and squawks. In recent evolutionary time, ravens have adapted themselves more and more to these open dry landscapes.

Soaring high above, red-tailed hawks (*Buteo jamaicensis*) are another constant avian feature of the landscape. A pair of red-tailed hawks might be observed hunting for prey from on high, soaring in wide circles, looking for a vulnerable rabbit, small bird, or reptile. Rock wrens (*Salpinctes obsoletus*) flit from between rock crevices, calling their loud dry trill, enlivening the quiet of a hot, sleepy afternoon.

As the day cools into evening and insects fly again, cliff swallows (*Hirundo pyrrhonota*) can be seen chasing them with erratic movements late in the waning light. The swallows build their nests of mud dabbed together into pottery-shaped houses

Black-tailed jackrabbit (*Lepus californicus*) frozen in place.
Photo by Stephen R. Jones

Ravens (*Corvus corax*) feeding on a carcass.
Photo by Cathy L. Cripps

Swallow nests clustered below a cliff overhang.
Yellowstone National Park Photo

on cliffs. They can be observed picking at mud in wet areas for this purpose. Common nighthawks (*Cordeiles minor*) soar above the shrublands on warm summer evenings, flying in loops and sounding their characteristic electric "pee-ent" call as they dive for insects in the quiet skies.

Seasoned hikers in these interesting semi-arid shrublands have learned by experience to tread carefully around members of the *Opuntia* genus, or prickly pears scattered among the rocks, grasses, and native shrubs. These cacti are indeed prickly, endowed by evolutionary forces to defend their edible, and often sizable, pads and "pears" by covering them with clusters of sharp spines. These formidable spines evolved as vestigial leaves and their bases produce tiny hairlike irritating barbed glochids. The succulent pads act as stems and serve to store life-giving water. *Opuntia* and other cacti glorify the arid land during the short flowering season with splashes of yellow and magenta red blossoms.

Throughout the western United States and elsewhere, cacti are used as a food source (after the spines are removed) for both domestic animals and humans. The fruits that form after the flowers fade are used in some parts of the region to make beautiful candies, jellies, and drinks.

The spiny pads and yellow flowers of the prickly pear cactus *Opuntia*. *Photos by Michael Kuo*

Another prickly plant that thrives in our xeric and semi-desert shrubland regions is the commonly occurring yucca or soapweed, *Yucca glauca*. Mashing the stems and roots of yucca plants produces a soapy lather, hence the common name. Its characteristic dense clusters of needle-sharp pointed leaves are conspicuous, especially when the glorious creamy white elongated clusters of bell-shaped flowers decorate the arid lands. In the spring those flowers are pollinated by very specialized yucca moths in a highly evolved mutualism whereby the moth pollinates the flowers and its larvae then feed on some of the developing seeds.

Members of the barberry family are found in semi-arid shrublands, adding refreshing colors to the subdued landscape. The thorny *Berberis fendleri* shown here brightens the landscape with its yellow clustered flowers and its attractive edible berries in the autumn, providing food for the local birds and animals as they prepare for winter. In human history, the berries have long been used as an acidic flavoring for food and for making jellies. The roots produce a nice yellow dye.

Significant areas of the semi-arid shrublands are covered with biological crusts, also called cryptogamic soils. The crusts are a consortium of

The tall sharp pointed leaves of the yucca plant in the semi-arid shrubland. *Photo by Loraine Yeatts*

Shrubs of barberry and the yellow flowers close up. *Photos by Loraine Yeatts*

Biotic soil crust primarily of lichen; a Utah sign advising hikers not to "bust the crust" by going off trail. *Photos by Cathy L. Cripps*

organisms living in quiet harmony. Lichens, a mainstay of these communities, pull carbon dioxide and nitrogen from the air and add them to the environment. The atmospheric nitrogen is processed by the cyanobacteria harbored by some lichens into forms that can be used by plants. This is similar to the process by which bacteria in the root nodules of legumes such as locoweed (*Astragalus*) "fix" nitrogen. Free-living algae join the crowd in some crusts, as do mosses and microscopic fungi. These fragile crusts serve to add nutrients to the soil and hold it in place. Disturbance and subsequent loss of these crusts has been linked to more frequent and serious dust storms in the Southern Rockies. Other threats to this environment include a loss of water resources due to diversion and influxes of invasive plants in disturbed areas.

The odors after rain in shrublands can be intoxicating, and fungal fruiting follows rain in arid environments. Here, water runs off fast through porous soils or gushes down canyons that soon turn dry again. The fungi have adapted to fruiting fast and then persisting. The so-called "desert decomposers" such as *Montagnea*, *Podaxis*, *Battarrea*, and *Tulostoma* all have tough, woody stems that hold their fruiting bodies in place over months and reach down to moisture deep in the soil. The mushroom-shaped *Podaxis* does not open its caps to the arid environment but remains closed. *Montagnea* on the other hand, opens and produces dry plates that detach and blow around in the wind, dispersing spores. Earthstars (*Geastrum*) and the puffball groups (*Calvatia*, *Bovista*, *Lycoperdon*) keep their spores enclosed inside a papery case. When *Bovista* fruiting bodies dry out, they detach themselves from the earth and roll around the open landscape like tumbleweeds.

Hunting for mushrooms in the semi-arid shrublands is more serendipity than intentional; meaning that fungi are few and far between and they are usually stumbled upon rather than actually sought out in these dry environments. But you can

Woody stalks of *Montagnea arenaria* and a *Tulostoma* species. *Photos by Robert Chapman, Cathy L. Cripps*

increase your chances by looking a few days after heavy rains drench the dry soils, especially in late summer and at times when the vegetation is low; your eye might catch a sight of some of the more persistent fruiting bodies. Near the edges of ditches and swales where moisture might collect, on the protected sides of rocky outcrops, and in the shady side of shrubs and bushes are all good places to examine for shrubland fungi.

A well-adapted small tree or shrub, *Quercus gambelii*, is common in the semiarid areas of the Southern Rocky Mountains. Gambel oaks, often called scrub oaks, support mycorrhizal fungi, those that live in a symbiotic relationship with the roots of the oak shrubs. Good examples are some members of the genera *Russula*, *Lactarius*, and *Amanita*. In the shade or protected areas under oak shrubs, puffballs, and earthstars, gasteroid fungi may be found a few days after heavy rains in the summer and early fall. If you venture into these dry lands, be sure to bring plenty of water!

Photo by Ed Lubow

Cercopemyces crocodilinus T. Baroni, B. Kropp, and V. Evenson

DESCRIPTION: CAP 3–9 cm across; convex, depressed slightly in center at maturity; white, with pale buff-brown base colors; surface distinctly covered with whitish warts or pyramidal scales; margin remaining inrolled and decorated with small warts to form a floccculose appendiculate margin. **GILLS** white; adnexed, not free, crowded, narrow. **STALK** white with brownish tints; short and stocky, 4–6 × 2–3 cm, more or less equal down to a distinct turnip-shaped base; top of the base decorated with small warts similar to those of the cap margin; whitish mycelium at base collects debris and soil. **FLESH** white, fairly solid, not staining at injury; odor mild. **SPORES** whitish in print.

ECOLOGY: Rare, solitary or gregarious, in humus near mountain mahogany, *Cercocarpus ledifolius* or *Cercocarpus montanus*, in semi-arid regions of the Rocky Mountains; early summer.

OBSERVATIONS: There are three known populations of this rare and, until recently, unrecognized fungus: one each in Colorado, Montana, and Utah. Although the populations are widely separated, the common habitat under mountain mahogany in high-elevation, xeric shrublands makes the collectors optimistic that more populations will be discovered. In both cases, collectors initially thought they had found an unusual *Amanita*. However, the narrowly attached gills and microscopic characters do not fit *Amanita*; extensive genetic work confirms that this is a previously undescribed genus.

Photos by Robert Chapman, O.K. Miller, Jr.

Podaxis pistillaris (Persoon) Morse

DESCRIPTION: FRUITING BODY 5–20 cm tall, consisting of an elongated cap and stalk, **CAP** narrowly elliptic, 2–10 cm long × 1.5–3 cm across; surface two-layered; outer surface shaggy, dull white with flattened buff-colored scales that eventually fall off exposing a smooth, brown, brittle inner surface; inner surface breaks up irregularly with age, revealing a dark brown to black, powdery, glebalike spore mass. **STALK** central; whitish; at first squamulose becoming smooth or somewhat striated; 3–10 cm long × 0.5–1 cm thick, equal to a small abruptly bulbous base; volva absent; texture firm and fibrous with a more or less woody cortex.

ECOLOGY: Single to scattered, growing in sandy arid soil, along roadsides and dry washes in desert shrublands; while not common, it is a remarkable find; developing in the spring through the fall after heavy rains; often persisting for months in inhospitable environments. *Podaxis pistillaris* is reported worldwide in many desert areas and occurs in xeric and semi-arid parts of the Rocky Mountains from Idaho to Arizona.

OBSERVATIONS: *Podaxis pistillaris* is a secotioid fungus, i.e., having evolved a cap that remains closed during its development in order to protect its fertile surfaces and spores from desiccation until they are mature enough for spore dispersal. Sometimes called the Desert Shaggy Mane because of its superficial resemblance to *Coprinus comatus* (Pg 61), genetic studies have not confirmed a close relationship. *Montagnea arenaria* (Pg 37) also occurs in sandy, desert areas and is similar by having a woody stalk but it arises from a volval basal cup and has a much reduced cap containing a crumpled gill-like fertile area.

Photo by Vera S. Evenson

Bovista pila Berkeley and M. A. Curtis

DESCRIPTION: SPORE CASE globose to subglobose, 2–7 cm across; at first with white, fuzzy surface, wearing off to expose inner skin that is papery thin, metallic bronzy purplish, smooth; an irregular apical pore or simple ragged tear eventually forms near the top, releasing the spore mass; sterile base absent; base attached to soil by a single cordlike extension. **SPORE MASS/GLEBA** at first white, then becoming deep purplish, powdery; odor and taste mild.

ECOLOGY: Scattered or in groups of several fruiting bodies, saprobic in soil and leaf debris, open woods, shrublands to montane regions of the Rocky Mountains. Found in late summer to fall.

OBSERVATIONS: The single strong strand that attaches these interesting puffballs to their growth substrate eventually breaks, releasing the tumbling puffballs to roll around on the surface of the semi-arid soils, blown by the ever-present breezes, releasing spores as they tumble. *Bovista plumbea* is a close relative, with bluish gray surfaces with age, but instead of a single cord of attachment to the substrate, *B. plumbea* has a clump of mycelial fibers at the base.

Photo by Cathy L. Cripps

Calvatia booniana A. H. Smith

DESCRIPTION: SPORE CASE very large, up to 50 cm across × 25 cm high; ovate to depressed globose; surface whitish, distinctly decorated at maturity with large, flat, angular scales; scales eventually flake off leaving a persistent layer that finally disintegrates to release the spores. **STERILE BASE** absent; spore case sits directly on soil surface, sometimes with a short basal attachment. **FLESH/SPORE MASS** white at first, slowly becoming yellowish olive to finally olive-brown and powdery; odor mild, currylike pungent in aging specimens; taste mild and pleasant.

ECOLOGY: Found in open grasslands from the high prairies to semidesert shrublands and lower foothills, this giant puffball is unique to western North America. This huge saprobe fruits from summer until fall, often gregarious and sometimes forming large fairy rings.

OBSERVATIONS: *booniana* named for Dr. William J. Boone of the College of Idaho who first showed this interesting local native to Dr. Alexander Smith, well-known mycologist. *Calvatia booniana* has been collected and eaten in many local areas since pioneer days and undoubtedly before that. Legend has it that settlers who built sod houses in the prairie lands used these and other large puffballs to fill holes in their dwellings to keep cold air out.

Photo by Robert Chapman

Geastrum schmidelii Vittadini

DESCRIPTION: FRUITING BODY small, as expanded 1–2.5 cm across × 1–2 cm high; dull grayish orange. **SPORE CASE** roughly globose, with distinctly pointed, raised apex; single torn pore at tip of apex, peristome distinctly delimited by brownish striations; spore case rests on short pedicel, up to 3 mm long at maturity. **RAYS** resulting from the splitting of the outer skin, separating into 7–10 rays; upper surface roughened and brownish; under surface with small accumulation of dirt and debris. **SPORE MASS** soon powdery brown with age; odor mild.

ECOLOGY: Terrestrial, often found in leafy debris of juniper in semi-arid regions, also in dry soil among prairie shrubs and grasses. Usually found in small groups, fruiting after summer rains, often persisting throughout the fall season.

OBSERVATIONS: *Geastrum* from *geo* meaning earth and *aster*, meaning star; the common name for these fascinating fungi is "earthstar." The star shape is the result of the outer layer of the nearly spherical maturing fruiting body splitting into rays that turn outward and in some species arch to raise the spore case higher into the air for better spore dispersal. In some species of earthstars called Barometer Earthstars, the rays are hygroscopic; that is, the rays open when moisture levels are high and close when they are low.

Photo by Robert Chapman

Mycenastrum corium (Guersent) Desvaux

DESCRIPTION: SPORE CASE globose, subglobose, to somewhat pear-shaped, often flattened; 4–7 cm high, 5–15 cm wide; two-layered; outer surface a firm, whitish felted layer that cracks into flat, blocklike patches that eventually fall away, revealing a thick, tough, persistent, dingy brownish inner layer; inner layer finally splits into lobes that sometimes become starlike and recurve. **NO STALK OR STERILE BASE**, but mycelial fibers often present. **SPORE MASS** (gleba) powdery, deep yellowish brown to dark reddish brown; odor mild or slightly unpleasant; taste mild.

ECOLOGY: Living up to its common name of the giant pasture puffball, *Mycenastrum corium* is found in open areas, horse pastures, and open sagebrush communities; scattered to gregarious, summer to fall; western North America, most common in the Rocky Mountains.

OBSERVATIONS: Large *Calvatias* such as *C. bovista* (Pg 27) or *C. booniana* (Pg 41) differ macroscopically from the giant pasture puffball by having warts or scales on the outer surfaces rather than the distinctive flat patches that develop from the felty two-layered skin of *Mycenastrum corium*. In addition, the genus *Mycenastrum* is unique in having tapering, thick-walled, pointy, thornlike capillitium (hyphae in the gleba of puffballs). A new subspecies, *M. corium* ssp. *ferrugineum* has been reported growing in an open space in Colorado near Denver. It has a distinctive rusty red gleba.

Photo by Robert Chapman

Myriostoma coliforme (Persoon) Corda

DESCRIPTION: FRUITING BODY more or less globose, 5–10 cm broad, at first encased in an outer multilayered covering that splits open from the top into rays that spread and often recurve, exposing the spore case. **SPORE CASE** is somewhat flattened and subglobose, 4–8 cm wide; inner peridium papery, pale silver-grayish and minutely roughened, distinctively featuring several porelike mouths on its surface; spore case is elevated on several short, slender, whitish columns. **RAYS** 4–8, arching and often recurved downward, not hygroscopic. **SPORE MASS** dark brown and powdery at maturity.

ECOLOGY: Solitary to gregarious, growing in sandy soil, widely distributed in many parts of the world in steppe and semi-arid regions; most of the reports from the Rocky Mountains region are from the states of New Mexico and Arizona, late summer into fall; never common.

OBSERVATIONS: This interesting rarity is one of our larger earthstars. Commonly called the salt-shaker earthstar or pepper pot because of its numerous porelike mouths; it is listed as protected in some parts of the world. See also *Geastrum schmidelii* (Pg 42).

Photo by Vera S. Evenson

Tulostoma fimbriatum Fries

DESCRIPTION: SPORE CASE subglobose, 1–2 cm across; outer surface at first a thickish sandy layer that gradually disappears revealing a tough, smooth membrane, pale grayish to pale brownish; lower third of case remains persistently coated with sand. **PORE MOUTH** central, barely raised, wrinkled, roughly circular, lacerate-torn. **STALK** 2–6 cm × 3–5 mm, tough, cylindrical, scurfy, often furrowed longitudinally as it dries, shrinking at point of attachment to spore case, leaving a distinct collar; stalk base expanding into a sandy small bulb. **FLESH** unchambered, fairly firm becoming powdery at maturity; reddish cinnamon; odor and taste mild. **SPORE MASS** cinnamon-brown.

ECOLOGY: Scattered to gregarious, often partially embedded in sandy, loose soil in semi-arid areas amid grasses and shrubs; spore case developing underground during moist weather and pushed upward on a long stalk so the spores can be dispersed into the air currents; fruiting in summer to fall.

OBSERVATIONS: Often called *Tulostoma fimbriatum* var. *campestre*. Known as stalked puffballs, these persistent fungi are often not discovered till fall or even winter when the shrubs and grasses have died back. *Tulostoma cretaceum* is another species found in xeric grasslands and gypsum soils; the base of its stalk is distinctly radicating and forking, but the collection has to be carefully excavated to show this interesting feature.

Photo by Melissa Kuo

Pycnoporus cinnabarinus (Jacquin) P. Karsten

DESCRIPTION: CAP 1.5–3 cm across; semicircular to kidney-shaped in outline; broadly convex or nearly flat; dry; finely hairy to suedelike when young, becoming bald or pocked with age; bright reddish orange when fresh, becoming dull orange with age; the margin often soft. **PORE SURFACE** with 2–4 round to angular or sometimes elongated pores per mm; bright reddish orange; occasionally extending onto the substrate below the cap; tubes to 5 mm deep. **STALK** absent. **FLESH** tough; reddish to pale orange; odor and taste not distinctive. **SPORE PRINT** white.

ECOLOGY: Saprobic on the deadwood of Gambel oak and other hardwoods in semi-arid habitats; growing alone, scattered, or gregariously; summer and fall.

OBSERVATIONS: The bright colors and habitat easily identify this species. *Pycnoporellus alboluteus* (Pg 202) is another bright orange Rocky Mountain polypore, but it is found at high elevations near melting snowbanks in spring.

Photo by Cathy L. Cripps

COTTONWOOD RIPARIAN

The headwaters of the great river systems of the Rocky Mountains begin quietly on snow-capped mountain peaks. Winter snow melts into alpine rivulets that flow and tumble into steep mountain streams. The streams converge to become the great rivers of the Rocky Mountains: the Yellowstone, Missouri, Colorado, Platte, and Snake, which flow east or west from their source along the Continental Divide. These waterways are the great connectors of our forest habitats running from the alpine to the prairie.

The word *riparian* (*ripa-*) refers to the vegetation along the banks of rivers and this could be extended to all waterways. This vegetation transitions from wet mosses in the alpine to willow wetlands in forests, all the way down to the grasslands and valleys where cottonwood groves line our major western rivers. Here, we focus on the broad bands of cool, shady cottonwoods that form an arching canopy over dense shrubs and crowded herbs along river and stream banks. The sound of babbling

Deeply furrowed bark of a large black cottonwood.
Photo by Cathy L. Cripps

brooks or the rush of river water joins with the cries and songs of birds to make the riparian a noisy and lively place for collecting mushrooms compared with the regal quiet of conifer forests.

Overstory trees include black cottonwoods (*Populus trichocarpa*) in the north, Fremont's cottonwood (*Populus deltoides* subspecies *wislizeni* = *P. fremontii*) in the south, narrowleaf cottonwood (*P. angustifolia*) and the plains cottonwood (*P. deltoides*). One of the largest plains cottonwoods is found in Boulder County, Colorado; it stands 105 feet tall and has a circumference of 35 feet. In spring, these trees are recognized by their excessive production of sticky, cottony catkins that fill the air like snow and eventually cover the ground. The bark of older cottonwoods is deeply furrowed and often accented with a touch of orange lichen.

Tucked below the cottonwoods is a diversity of small trees and shrubs, most consistently willow (*Salix* spp.) and alder (*Alnus* spp.), but also including chokecherry, red-osier dogwood, hawthorn, box elder, river birch, and wild rose. These woody plants can thicken the understory to an almost impenetrable tangle. Patches of horsetails and cattails shoot up in wet areas, and stinging nettles, thimbleberry, Solomon-seal, and violets are among the plants that

Cottony catkins covering the ground, and a dense riparian understory in autumn. *Photos by Cathy L. Cripps*

Riparian understory of chokecherry (*Prunus virginiana*); horsetails (*Equisetum*) with *Psathyrella*, and violets (*Viola canadensis*). *Photos by Cathy L. Cripps*

proliferate along banks. In sandy areas, open cottonwood stands support sparse vegetation and the soil can become dry and hard in summer.

This diversity of plant life draws an array of nesting and migratory birds, especially in spring. In parts of the Rocky Mountains, over 50 percent of the regional bird species nest in the riparian. Ducks, Canada geese, pelicans, great blue herons, kingfishers, American dippers, egrets, and cranes all wade, bob, or dive in the mountain streams and larger rivers. Sandhill cranes follow watercourses on their annual migrations. Eagles and osprey build their nests in the cottonwoods and cruise the rivers for fishy sustenance. In winter, eagles can be easily observed sitting in leafless cottonwoods waiting for prey. Along the banks, woodpeckers, flickers, chickadees,

Fishing: Osprey with trout on cottonwood and great blue heron with minnow. *Photos by Andy Hogg*

Bald eagles nesting in a cottonwood. *Photo by Andy Hogg*

Red-winged blackbird on cattails. *Photo by Cyndi Smith*

warblers, goldfinches, and yellow-headed and red-winged blackbirds seek the deeper recesses of the riparian habitats.

Mammals crisscross the riparian seeking food and water; paw- or hoofprints on sandbars are a record of their nightly adventures. Rabbits, deer, and raccoons make this leafy environment their home and occasionally a bear will pass through. Closer to the water, beaver, muskrats, and river otters swim, play, and sun themselves along the banks. Beavers (*Castor canadensis*) seriously alter stream flows by engineering dams of cottonwood and aspen logs and branches. Trout rise during periodic insect hatches or swim lazily beneath the water's surface. In the late 1800s, Tamarisk (*Tamarix* spp.) and Russian olive trees (*Elaeagnus angustifolia*) were planted to prevent erosion of riverbanks, and these fast-reproducing hardy exotics took over in many riparian areas of the West. Unfortunately, they do not provide the food or habitat important for native animals as do the cottonwoods. Other threats to these habitats are livestock grazing, dams, and reservoirs, which alter natural stream ecology.

Cool, shady riparian habitats are often overlooked as places to hunt mushrooms but they offer a diversity of fungal forms. Spring is an exciting time, when yellow morels and oyster mushrooms can sometimes be gathered by the basketful. River runners are known to take advantage of riparian areas that are hard to get to by accessing them by boat. Crawling through the tangled understory is another technique for finding overlooked morels and oysters when there is competition. The best collecting times are usually in May, just before and after high water during spring runoff; this is often when domestic lilacs bloom in yards in the region. When conifer

Beavers gnaw down cottonwoods and use them to build dams. *Photos by Andy Hogg, Cathy L. Cripps*

Riparian inhabitants: brown trout and a cottontail rabbit. *Photos by Cathy L. Cripps*

Yellow morels show color variation. *Photo by Cathy L. Cripps*

Pale young oyster mushrooms on a cottonwood. *Photo by Cathy L. Cripps*

forests are dry in the Rocky Mountains, riparian areas with their mix of cottonwoods, willows, and alders can hold moisture late into summer.

Woody riparian plant life is deciduous and trees and shrubs all lose their leaves in fall. This pile of plant matter would build up excessively except for the myriad of riparian fungi that break it down and recycle it back into the soil. Miniscule Mycenas, tiny Tubarias, petite cup fungi, and an assortment of polypores not found in conifer forests do the work of beneficial decay. The saprophytic spine fungus *Hericium coralloides* fruits on cottonwood logs and is a riparian delicacy that is often sought for the table. Late in the year (even into October and November) the edible *Tricholoma populinum* can be gathered in quantity under cottonwoods if you can beat the deer to them.

Access to riparian habitats is not always easy in the Rocky Mountains because of private property and/or prohibitive fences and signage. Wise mushroomers check areas open to fishing and boating at lower elevations, including those within national forests. In the north, the banks of the Missouri and Yellowstone Rivers support large cottonwood groves, and in the Southern Rockies, the banks of the Platte, Snake, and Colorado Rivers have extensive riparian vegetation worth visiting. A boat is helpful to access some cottonwood stands along the larger rivers. The smaller rivers and streams can also be productive depending on their pristine quality. Banks infested with tamarisk and Russian olive are not good areas for collecting. In any case, a walk in the cool, shady riparian in summer is a delight of surprises, whether fowl, furry, or fungal.

Photo by Cathy L. Cripps

Amanita populiphila Tulloss and E. Moses

DESCRIPTION: CAP 6–10 cm across, convex, then almost flat; pale orange, pale orange-brown or pale orange-gray; smooth in center and deeply grooved at the margin; covered with patches of whitish cream tissue that can be sparse, abundant, or absent. **GILLS** free, crowded, broad; cream or pale orange, graying slightly with age, especially at the edges. **STALK** 7–15 × 0.5–2.5 cm, tall and slender, slightly larger at base but without a bulb; white; surface covered with small flecks, sticky; ring absent; base enclosed in white fragile tissue that forms a loose sac (volva). **FLESH** white, stem stuffed; odor not distinctive. **SPORE PRINT** white.

ECOLOGY: Mycorrhizal; found in dryer riparian areas near cottonwoods, aspen, and willows, often in more open grassy spots; fruiting in late spring and early summer; known from Montana, Wyoming, Colorado, Idaho, and Kansas.

OBSERVATIONS: *populiphila* for its hosts; related to the European *Amanita fulva.* This *Amanita* lacks a ring and is distinguished from others in the Vaginatae group by its habitat and pale orange colors. This species has been reported fruiting in great numbers under cottonwoods. *Amanita barrowsii* nom. prov. is similar but its volva is a striking orange on its inside surface.

Photo by Cathy L. Cripps

Laccaria montana Singer

DESCRIPTION: CAP 1–2 cm across, shallow convex, sometimes with a dimple in the center; dark orange or red-brown when fresh, drying pale orange-brown; slightly translucent, surface smooth but with a lined margin. **GILLS** broadly attached; well-separated (only 16–20); pink; rather thick. **STALK** 2–5 cm × 2–3 mm wide, thin, equal, fibrous; orange-brown, with white mycelium at the very base. **FLESH** pale orange; rather tough-fibrous in stalk; odor indistinct. **SPORE PRINT** white; spores round and spiny, 4 per basidium.

ECOLOGY: Fruiting in groups or clusters in wet riparian areas along streams in the Rocky Mountains; mycorrhizal with aspen, cottonwood, and willows, and with other hosts in boreal and northern habitats. Also known from the alpine.

OBSERVATIONS: *montana* for its mountain distribution; this species can be confused with *L. laccata* but is smaller and more delicate and is technically separated by its larger spores (up to 14 µm). *L. montana* also is more common in wet areas, whether it be in riparian, boreal, or alpine habitats. The similar *L. pumila* also occurs in wet habitats with woody hosts but it has 2 spores per basidium.

Photo by Ed Barge

Lactarius zonarius var. *riparius* Hesler and A. H. Smith

DESCRIPTION: CAP 5–8 cm across, depressed in the center and margin strongly rolled under, thick-fleshed; cream with orange-brown colors developing with age or on handling, some with faint or distinct concentric zones; surface smooth, sticky, with soil adhering; edge of cap slightly hairy at first but soon smooth. **GILLS** run slightly down the stalk, crowded; cream; exuding white milk that does not change color when cut but stains gills brownish. **STALK** short and stout, 2–4 × 1.5–2.5 cm, pointed at the base with soil adhering; cream color, without spots. **FLESH** very firm, cream; odor sweet, fruity; taste slowly hot. **SPORE PRINT** light.

ECOLOGY: In dense clusters half-buried in soil near streams in riparian areas, under willows and cottonwoods, especially in spring, but also in fall; reported as mycorrhizal with black and plains cottonwoods in the Rockies but originally described by Alex Smith from Michigan under brush in riparian areas.

OBSERVATIONS: *zonarius* var. *riparius* for its faintly zoned cap and riparian habitat. The cream-colored fruiting bodies eventually turn orange-brown and become more zonate with age. This large, firm, robust species can look similar to other riparian fungi such as *Paxillus* (brown spores) and *Tricholoma populinum* (no milk and white-spored). While large, meaty, and abundant, this *Lactarius* is not for eating because of the acrid taste.

Photo by Vera S. Evenson

Mycena acicula (Schaeffer) P. Kummer

DESCRIPTION: CAP 4–10 mm broad; convex with low umbo when young, becoming broadly bell-shaped with age; smooth to faintly striated; distinctly colored scarlet to coral-red, slowly fading to bright orangish yellow; margin sharp and undulating, yellowish. **GILLS** whitish to pale yellowish; attached-adnexed, broad, subdistant; edges smooth. **STALK** 1–5 cm long, up to 1 mm across; cylindric, hollow; light yellow to lemon yellow; without an annulus; base with white hairs. **FLESH** yellowish; thin, somewhat brittle; odor and taste not distinctive. **SPORE PRINT** white.

ECOLOGY: Saprobic; gregarious, sometimes in small clusters; growing in damp places along streams, at times in floodplains, within leaf litter and hardwood debris; fruiting in spring and early summer, occasionally in wet weather in the fall.

OBSERVATIONS: *acicula* for its pinlike appearance. This lovely, brightly colored mushroom is not common in its habitats but is usually noticed because of its distinctive brilliant colors. It is widely distributed in Colorado, occasionally in Montana and the Pacific Northwest; also known from Europe. *Mycena adonis* is similar but is typically found under conifers.

Photo by Cathy L. Cripps

Pleurotus pulmonarius (Fries) Quélet

DESCRIPTION: CAP 4–20 cm across and larger, shell-shaped to almost flat with age, arranged in overlapping clusters on wood; pale buff, pale brown, brown, gray-brown; smooth, dry to greasy; edge rolled under when young, sometimes wavy with age. **GILLS** run down the stalk; at first as shallow ridges, later becoming broad and crowded; white or cream. **STALK** absent or short and off to one side, 1–4 cm long × 1–2 cm wide, merging with those from other caps; white; surface ridged, often with a few bristles. **FLESH** white; firm when young, tough in stalk; odor fragrant, fungoid, pleasant. **SPORE PRINT** lilac-gray (in a heavy spore print).

ECOLOGY: Saprobic; typically forming large clusters on cottonwood stumps, logs, and standing trees in riparian areas; fruiting in early spring and again in late fall when temperatures cool; known from the Rocky Mountains, but also on the East and West coasts; reported on several species of cottonwoods and conifers.

OBSERVATIONS: *pulmonarius* for its lung-shaped fruiting bodies. This common oyster of the Rocky Mountain riparian is more highly colored (brown or gray-brown) than the whitish aspen oyster *P. populinus* (Pg 110), which also lacks a lavender spore print. Both are edible and prized if you can beat the insects to them. Careful pruning helps ensure maturation of younger caps. The related *P. ostreatus* does not occur or is rare in the Rockies.

Photo by Cathy L. Cripps

Tricholoma argyraceum (Bulliard) Gillet

DESCRIPTION: CAP 2.5–5 cm, conic-convex or almost flat, with a small central bump and turned down margin; white with gray-brown radial fibrils that can wear away; edge can be a bit ragged looking. **GILLS** attached, rather broad, white, edges eroded, staining yellow with age. **STALK** 2–6 × 1 cm, equal, some curved; white with a few fibrils. **FLESH** white, brittle, stem stuffed, staining yellow with age; odor of fresh meal (floury) or blossoms (flowery). **SPORE PRINT** white.

ECOLOGY: In scattered groups in riparian areas under cottonwoods in late spring, summer, and fall in the Rockies; mycorrhizal with other *Populus* species in the same habitats in Europe but not well-known in North America.

OBSERVATIONS: *argyraceum* for its silvery appearance, but it can appear more whitish with age. There are several small gray Tricholomas in the Rocky Mountains; it is interesting that each appears to associate with certain tree species. *Tricholoma scalpturatum* also turns yellow with age but has a flatter cap and occurs more with aspen; *T. cingulatum* has a definite ring and is with willows. Other gray Tricholomas such as *T. terreum*, *T. myomyces*, and *T. moseri* prefer conifers. None is considered edible; some Tricholomas are toxic.

Photo by Cathy L. Cripps

Tricholoma populinum J. E. Lange

DESCRIPTION: CAP 7–12 cm across, irregular, slightly domed or flat with a wavy uplifted margin; smooth, sticky, covered with adhering soil; dingy pinkish buff to coppery red-brown in center, paler at the cap edge. **GILLS** attached, crowded, broad; dingy white to pale gray-brown with a pink tint. **STALK** big, fleshy, 4–10 × 2.5–5 cm, broadest at the top and narrowing to a point; whitish but streaked with brown at the base, which is covered with adhering soil; ring absent. **FLESH** thick, firm, white; odor of fresh meal. **SPORE PRINT** white.

ECOLOGY: Occurring in very dense clusters of overlapping caps often in hard-packed soil beneath black cottonwoods (*Populus trichocarpa*) in the northern part of the Rocky Mountain region or with *P. deltoides* in the Southern Rockies. In Europe, it is mycorrhizal with other *Populus* species. One of the last mushrooms to fruit in fall, it appears as late as September and October.

OBSERVATIONS: *populinum* for its mycorrhizal hosts. This large *Tricholoma* is recognized by the dense clusters, which can appear as large "mush lumps" pushing up clods of soil in dry areas. It is an edible species that has been used as a food source by the Salish Indian peoples of British Columbia as well as Pueblo Indians near Taos, New Mexico. The collection in the photo is a molecular match to European specimens of *P. populinum*. *Tricholoma fulvimarginatum* is similar but has gill edges that turn brown; it is also large and found under cottonwood.

Photo by Cathy L. Cripps

Coprinopsis variegata (Peck) Redhead, Vilgalys, and Moncalvo

DESCRIPTION: CAP 3–5 cm across by 2–3 cm high; egg-shaped at first, then bell-shaped, with a pleated margin; cream to pale gray-brown; covered with felty patches of cream-colored tissue when young; edges turn gray then black as the cap autodigests. **GILLS** free, very crowded; grayish then black with white edges, turning to ink when mature. **STALK** 7–10(15) × 1 cm, somewhat long, equal; white; surface smooth but with a few patches of tissue, hollow; some with a slight ring zone. **FLESH** white but turning to blackish ink beginning at the cap edge; odor not distinctive. **SPORE PRINT** black.

ECOLOGY: In large clusters in riparian areas especially in leafy hardwood debris in spring and early summer. Once considered a saprobic species of eastern hardwood forests, it is also reported from Montana, Wyoming, and Alberta.

OBSERVATIONS: *variegata* for the patches of felty tissue. This species looks similar to the well-known edible shaggy mane *Coprinus comatus* (Pg 61) but DNA analysis reveals it to be only distantly related. Mycologist Orson Miller reports this species from riparian habitats in spring in western states. It is not recommended as an edible as there are reports of digestive upsets and reactions after ingestion. *Coprinus quadrifidus* is similar but has a woody stalk.

Photo by Cathy L. Cripps

Coprinus comatus (O. F. Müller) Persoon

DESCRIPTION: CAP 5–15 cm tall by 2–4 cm wide; egg-shaped when young, becoming bell-shaped; covered with shaggy white scales over a white ground color; scales darkening to brownish mostly at the top of the cap; margin shaggy, turning pinkish, then black as the cap autodigests (turns to ink) from the edge upward. **GILLS** free, very crowded; white, then blackening. **STALK** 8–20 × 1–2.5 cm, equal, smooth; white; sometimes with a slight movable ring (*which has slipped to the base in the photo*). **FLESH** fragile, white, hollow in stipe; odor not distinctive. **SPORES** black.

ECOLOGY: Single or in clustered groups, often in disturbed soil, in grass, in logged-over forest areas, in the riparian, along roadsides, in gardens and mulch beds, from low to high elevations; sometimes coming up through hard-packed soil and even asphalt; fruiting in summer and fall; saprobic on buried organic material.

OBSERVATIONS: *comatus* for its common name, "shaggy mane." Considered a good edible when young before it turns black, it needs to be cooked up quickly. Some people have reported GI distress from eating *C. comatus*, whether consumed with alcohol or not. *Coprinopsis atramentaria* is grayer with a smoother (or finely scaly) cap and grows in dense clusters at the base of trees. This "alcohol-inky cap" contains coprine, which causes a negative reaction when consumed with alcohol; it is not recommended in any case. Also compare with the inedible *Coprinopsis variegata* (Pg 60) and *Chlorophyllum molybdites*; the latter looks similar, is found in grass, does not autodigest, gives a green spore print, and has caused numerous poisonings in Colorado and elsewhere.

Photo by Cathy L. Cripps

Psathyrella spadicea (P. Kummer) Singer

DESCRIPTION: CAP 5–9 cm across, shallow convex with a low knob in the center and a wavy margin; smooth, greasy; dark to milk chocolate brown, drying lighter to a clay-buff; sometimes with bits of white tissue on cap edge, which is faintly lined. **GILLS** attached and pulling away, broad; milk-coffee color, then brown or red-brown. **STALK** 5–7 × 0.5–1.2 cm, mostly equal, undulating; surface minutely powdery; white; fragile, hollow. **FLESH** white; odor faint. **SPORE PRINT** reddish to red-brown.

ECOLOGY: This decomposer occurs in large clusters on dead, often buried, cottonwood logs in early spring. It can fruit prolifically at times. It is broadly distributed in riparian areas of North America.

OBSERVATIONS: *spadicea* for the red-brown or date color of the cap. There are several Psathyrellas that occur in riparian areas in the Rocky Mountains but most are smaller. Smith (1979) mentions quite a few species and gives the technical details. *P. uliginicola* (Pg 118) is another large species but is more likely under aspen; it is stouter with a pale gray pleated cap. Neither is edible.

Photo by Cathy L. Cripps

Stropharia riparia A. H. Smith

DESCRIPTION: CAP 4–6 cm across, shallow convex, some with a low knob, smooth, greasy; pale yellow, becoming more yellow-brown in center; margin hung with pieces of white tissue. **GILLS** narrowly attached; gray with a slight lavender tint. **STALK** 6–8 × 0.5–0.8 cm, rather long, equal, slightly narrowing toward base; pale yellow-white, more brown at base, smooth with a few fibrils; slight fibrous ragged ring or at least a ring zone. **FLESH** white but yellow under cuticle and orange in the stipe base; odor fungoid or faint. **SPORE PRINT** purple-brown.

ECOLOGY: Scattered in riparian areas on dead cottonwood or aspen in spring and summer in the Rocky Mountains and the Pacific Northwest. This saprobic species is consistently associated with *Populus* wood.

OBSERVATIONS: *riparia* for its habitat. This species has been considered as a synonym of *Leratiomyces percevalii* (Berkeley and Broome) Bridge and Spooner, known from wood chips in European parks. The taxa are morphologically close but until molecular work is completed we choose to keep the two separate. A. H. Smith (1979) restricted his species to aspen and cottonwood habitats along streams. Most *Stropharia* species are not on wood or buried wood. Not an edible species.

Photo by Cathy L. Cripps

Hericium coralloides (Scopoli) Persoon

DESCRIPTION: FRUITING BODIES small to large, 12–25 cm across; coralloid, composed of numerous, repeatedly dividing, irregular slender branches coming from a base on one side; branches covered with pendant iciclelike spines along the lower side that are 1–6 mm long; branches and spines white to cream, more yellow with age. **FLESH** white, somewhat rubbery; odor mild to sweet. **SPORES** white, amyloid.

ECOLOGY: A white-rot fungus found on hardwood logs such as those of cottonwoods when they are devoid of bark; usually in moist understory situations along streams; rather rare but reasonably common at times during summer. It is known from both North America and Europe.

OBSERVATIONS: *coralloides* for its shape. The names of the various spine fungi in North America have been confused. Currently, *H. coralloides* is considered synonymous with *H. ramosum*, and these are separated from *H. abietinum*, which is on conifers, and *H. erinaceus*, which is not branched. All *Hericium* species are considered excellent edibles and some are also purported to have medicinal properties.

Photo by Cathy L. Cripps

Morchella esculentoides M. Kuo, Dewsbury, Moncalvo, and S. L. Stephenson

DESCRIPTION: CAP 5–10 cm tall by 3–6 cm wide, conic with a broadly rounded top; covered with irregular pits and ridges that are not in regular rows; grayish at first becoming yellow-brown when mature; margin attaches to the stalk; ridges light-colored and pits darker. **STALK** 4–8 × 1.5–3.5 cm, mostly constricted in the middle; cream; surface covered with small bumps; puckered at the base. **FLESH** cream; rather rubbery but brittle; whole fruiting body hollow; odor not distinctive. **SPORE PRINT** yellow or orange.

ECOLOGY: In riparian areas near cottonwoods (*Populus trichocarpa* and *P. deltoides*) and with *P. angustifolia* and *P. balsamifera* in the Southern Rockies. Often on calcareous soil as indicated by junipers in addition to the host trees. In the Rocky Mountains, yellow morel season begins in May if rains are sufficient and ends sometime in June. Blooming lilacs in the region are a good indicator the season has begun. *M. esculentoides* occurs elsewhere in North America and is thought to be loosely mycorrhizal with its host trees.

OBSERVATIONS: *esculentoides* for its edibility; this is the common yellow morel of North America previously called *M. esculenta*; another name is now *M. americana*. Yellow morels are eminently edible but should be cooked well and should not be eaten raw or when old (check the odor). At times, fruiting can be prolific in the Rocky Mountains and baskets can easily be filled. The similar *Verpa bohemica* (Pg 66) has a cap that is not attached at the margin and hangs free; it is not recommended as an edible.

Photo by Ed Barge

Verpa bohemica (Krombholz) J. Schröter

DESCRIPTION: CAP up to 4 cm wide × 5 cm tall, bell-shaped, covered with vertical wrinkles, with a few cross walls; yellow-brown, brown; margin hangs free, only attached to the stalk at the top; underside whitish. **STALK** up to 15 cm long and 5 cm wide, gradually larger toward the base; surface rough or bumpy; cream. **FLESH** pale, brittle, stuffed or hollow in stalk; odor not distinctive. **SPORES** light in color and very large.

ECOLOGY: Fruiting in troops in riparian habitats and also under aspen in the Rocky Mountains, and known throughout North America; fruiting on wet ground in spring from late May to early July just before or at the same time the true morels are fruiting under cottonwoods.

OBSERVATIONS: The first report of this species included a location in Bohemia. Verpas differ from true morels (*Morchella* spp.) (Pg 65) in that the edge of the cap hangs free. It is important to know the difference since *V. bohemica* can cause a negative reaction in some people especially if eaten raw, in quantity, or over consecutive days; it is best avoided.

MONTANE
Ponderosa Pine Forests
Douglas Fir Forests
Aspen Forests
Lodgepole Pine Forests
Burned Ground

Photo by Michael Kuo

PONDEROSA PINE FORESTS

The word *ponderosa* brings to mind "ponderous or large." That is just what ponderosa pines are: large native evergreen trees that form huge forests in western North America; the original scientific report described their timber as "very heavy"; their influence as a climax tree in the montane zone is indeed massive.

The state tree of Montana, *Pinus ponderosa*, is the most widely distributed pine in western North America with its natural range extending from British Columbia to Mexico and from the Pacific coastal region east to Nebraska. In the Rocky Mountains, ponderosa pines in nearly pure stands dominate many of the lower foothills, especially south-facing slopes, sometimes gently blending into the prairie grasslands down low and extending up in elevation to the pinyon-juniper arid shrublands. At even higher elevations in the Southern Rockies, up to 9,000 feet or more, they intermingle with lodgepole pines and aspen.

Ponderosa pines near Avon, Montana stretching into the valley.
Photo by Vera S. Evenson

Ponderosa pines have open canopies and are favorite spots for picnics and a nap in their shade. As these attractive pines mature, they gradually shed their lower branches so one can walk right up and inspect them closely.

Their unique bark is thick, scalelike, and orangey brown at maturity, decorated by deep blackish crevices. The mature trees give off a nice odor of vanilla when the sun warms the bark. The green ponderosa pine needles are typically flexuous and long, from 5–10 inches, growing in bundles of two or three, depending on the subspecies found throughout the range of this pine. Indigenous peoples have traditionally used these plentiful needles to weave long-lasting and beautiful baskets.

Parklike open stands are characteristic of these beautiful pine trees, but a more sober cause of the openness in a healthy ponderosa stand is wildfire. Historically, natural fires from sources such as lightning strikes burn through ponderosa groves quickly without igniting the mature trees that have evolved their characteristic protective thick bark. The fires clear out the undercover and ladder fuels, thus maintaining the sunny openness characteristic of ponderosa pine habitats.

The cones from ponderosa pines, characteristically warm brown and broadly shaped at maturity, are life-sustaining sources of food for native animals and birds. The Abert squirrel, *Sciurus aberti*, lives exclusively in ponderosa pine forests, using

Close up showing the deep crevices in the bark.
Photo by Vera S. Evenson

Long needles of ponderosa pine are in bundles of 2 to 3.
Photo by Vera S. Evenson

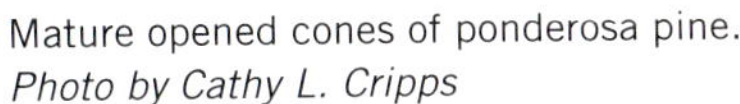

Mature opened cones of ponderosa pine.
Photo by Cathy L. Cripps

The Abert squirrel is recognized by its tasseled ears.
Photo by Michael Kuo

their branches for nesting and their seeds, buds, and cones as major parts of its food supply. Since Abert squirrels, commonly called "tassel-eared squirrels," do not make food caches, they are active year around. In addition to foraging within the pine trees themselves, these native squirrels harvest both aboveground mushrooms and hypogeous fungi that live in essential and mutualistic associations with these important native trees.

The open habitat under ponderosa pines allows many sun-loving plants to thrive there. The flowers of mountain ball cactus, *Pediocactus simpsonii*, brighten the semi-arid soils under pines in the springtime. A characteristic plant growing near ponderosa pines is *Grindelia squarrosa*, known locally as "gumweed." It has been traditionally used by Great Plains natives as a medicinal herb.

Prickly poppies, *Argemone polyanthemos*, flower in midsummer to late summer in many foothills habitats with ponderosa pine nearby. Their leaves are thorny and the stems exude a white latex. Seeds of these distinctive flowers are a source of nutritious oil for quail and other native birds.

Both mountain and western bluebirds inhabit ponderosa pine forests during the nesting season in various regions of the Rocky Mountains. They nest in holes in mature ponderosa pines, often excavated originally by local woodpeckers. The arrival of the bluebirds in the spring adds a special flash of sky blue to the green pines.

Pygmy nuthatches (*Sitta pygmaea*) are characteristic birds in ponderosa pine forests. They excavate

The brilliant colors of the mountain ball cactus in flower.
Photo by Vera S. Evenson

Ponderosa with gumweed *Grindelia squarrosa* and the flowers close up. *Photos by Vera S. Evenson, Loraine Yeatts*

nesting holes in trunks of the mature trees and live on pine seeds and insects they energetically pry from the irregular, cracked bark of the trees. During the winter these tiny nuthatches gather in large groups, sometimes as many as 100 individuals, to stay warm together inside the standing trunks of the pine trees. The pygmy nuthatch's communal living lends a happy atmosphere to spring and summer as their cheerful twitters and amusing acrobatics enliven the forest.

The top animal predator in ponderosa pine forests is the mountain lion, *Puma concolor*. Its main prey is the resident mule deer. One large deer can sustain a lion and her cub for several days. Other prey of this stealthy cougar includes native rabbits, squirrels, and even porcupines.

Prickly poppies in ponderosa pine and close up. *Photos by Vera S. Evenson*

Mountain bluebirds (shown) and western bluebirds are inhabitants of ponderosa pine forests. *Photo by Andy Hogg*

Pygmy nuthatches often huddle together inside the pines to keep warm. *Photo by Joseph Mahoney*

Mountain lion hidden in the branches of a ponderosa pine. *Colorado Division of Wildlife Photo*

Mule deer, *Odocoileus hemionus*, are very common inhabitants of pine forests from the foothills to the montane areas of large parts of the Rocky Mountains. Their large mulelike ears give them their common name. White-tailed deer, which also can be found in many areas of the Rockies, have smaller ears and distinctive white tails that they flag to communicate danger to their herd members and fawns. Both of these types of deer breed and shelter within forests. They are browsers, feeding on grasses and herbs in spring and summer and depending upon a wide variety of shrubs and even pine needles during the off season. It is well known among mushroom enthusiasts that mule deer also relish mushrooms, both above- and belowground species.

Mule deer posing in front of a ponderosa forest. *Photo by Stephen R. Jones*

Throughout the vast areas of ponderosa pine forests in so many parts of the Rocky Mountains, the vital association between the trees and mushrooms and other fungi is an absolutely essential part of the ecosystem. Many native plants, including most trees in our forests, share an amazing mycorrhizal relationship whereby the mycelium of the fungi and the rootlets of the trees are intimately connected in order to share minerals, carbon sources, water, and other compounds that benefit both types of organisms (Pg 6). A great variety of mushrooms live in

Collecting *Boletus barrowsii* under ponderosa pine. *Photo by Vera S. Evenson*

this close association with ponderosa pines; some of the better known genera are *Russula*, *Lactarius*, *Boletus*, *Suillus*, *Hygrophorus*, *Hebeloma*, *Amanita*, and *Tricholoma*, many species of which are featured in this book. *Boletus barrowsii* and *Neolentinus ponderosus* produce large meaty fruiting bodies in fall; they are both edible and often one fruiting body of either can be enough for several meals.

Other nonmycorrhizal fungi are involved in another essential role in the ponderosa pine habitats, that of breaking down and recycling the dead trees, cones, needles, and other woody debris that occur as the result of the natural growth and death of the trees. These fungal recyclers and rotters comprise several main genera commonly found in conifer forests, such as *Fomitopsis*, *Xeromphalina*, *Porodaedalea* (*Phellinus*), *Marasmius*, *Neolentinus*, and others.

Photo by Robert M. Chapman

Amanita sp.

DESCRIPTION: CAP 5–14 cm across, at first egg-shaped and covered with a thick white universal veil; upon expansion becoming hemispheric to convex with incurved margin; margin striated; caps peachy red, orange-ocher, to bright salmon-orange, centers often discoloring to pale ocher; smooth, viscid, with occasional large whitish patch of universal veil tissue adhering to surface. **GILLS** close to crowded, free, but slightly attached when young; moderately narrow; pale pinkish orange, edges yellowish. **STALK** 6–14 × 1–2.5 cm, cylindric to enlarged toward base, then pointed at very base; whitish with tiny pale orangish scales. **PARTIAL VEIL** copious, skirtlike, membranous; pale orange to yellow-orange; firmly attached to upper stalk, often collapsing. **UNIVERSAL VEIL** distinctively large as a volval cup, extending up to one-third the length of stalk; white; membranous, with a free margin that is often split, persistently attached at base. **FLESH** pale yellowish buff with layer of bright yellow just below cap cuticle; not staining at injury; odor mild and fruity and taste not recorded. **SPORES** pale yellowish in print.

ECOLOGY: This attractive *Amanita* is mycorrhizal under ponderosa pine in Arizona, New Mexico, parts of California, and northern Mexico; scattered to gregarious, often in arcs, fruiting after summer and early fall rains.

OBSERVATIONS: Often called Caesar's *Amanita* because of its superficial resemblance to the European *Amanita caesarea*; however, it differs in coloration and microscopically from that famous mushroom. A local provisional name is *A. cochiseana*. Although collected as an edible mushroom locally, caution is advised because of the dangerous reputation of the Amanitas.

Photo by Michael Kuo

Gymnopus perforans (Hoffmann) Antonín and Noordeloos

DESCRIPTION: CAP 4–17 mm across; convex, becoming broadly convex or broadly bell-shaped but often developing a central depression and/or bump; dry; bald; sometimes becoming slightly wrinkled or ribbed; pale brown when young, but often fading to buff or nearly white. **GILLS** narrowly or broadly attached to the stalk or occasionally attached by means of a "collar" that encircles the stalk; distant or nearly so; buff to whitish. **STALK** up to 40 mm long; up to 1 mm thick; equal; dry; finely hairy to finely velvety (more so toward the base); pale at the apex, but brown to dark brown, reddish brown, or nearly black below. **FLESH** thin; insubstantial; whitish to brownish; odor usually unpleasant (reminiscent of rotten cabbage) but sometimes faint or lacking and taste similar or not distinctive. **SPORE PRINT** white.

ECOLOGY: Saprobic on the fallen needles of conifers; growing scattered or gregariously; summer and fall.

OBSERVATIONS: Formerly known as *Micromphale perforans*, this tiny but stinky species is often overlooked. It is a regular in conifer forests throughout the Rocky Mountains.

Photo by Vera S. Evenson

Hygrophorus speciosus Peck

DESCRIPTION: CAP 2–5 cm across, convex, often with a low umbo; bright reddish orange to yellowish; smooth, viscid to glutinous; margins remaining inrolled and whitish. **GILLS** whitish with yellow tints; decurrent, subdistant, rather narrow, waxy; edges thick and often yellowish. **STALK** long, 4–10 cm × 0.5–1 cm across, equal or slightly wider toward base; white above, lower two-thirds coated with dull orange slime veil remnants. **FLESH** white to yellowish, firm, not staining; odor and taste mild. **SPORE PRINT** white.

ECOLOGY: Gregarious, sometimes subcaespitose, in needle duff and moist areas; mycorrhizal with conifers including larch, widely distributed in North America. *Hygrophorus speciosus* is associated with pines including ponderosa pines in moist seasons in the Southern and Northern Rocky Mountains; summer and fall.

OBSERVATIONS: This colorful *Hygrophorus* was first assumed to be found only with larch in moist areas and bogs but is now known to occur more broadly with other trees including pine and fir. *Hygrophorus hypothejus* is similar, but the glutinous cap surface is yellowish or olivaceous brown and it is always associated with pines.

Photo by Michael Kuo

Leucopaxillus albissimus (Peck) Singer

DESCRIPTION: CAP 3–20 cm across; convex with an inrolled margin when young, becoming broadly convex, flat, or shallowly depressed; dry; bald or very finely velvety; white, pinkish, buff, pale brownish, or pale tan; often darker with age and/or darker toward the center. **GILLS** attached to the stalk or beginning to run down it; separable from the cap as a layer; close, narrow; whitish to dirty yellowish. **STALK** 3–8 cm long; up to 3 cm thick; with a somewhat swollen base when young, but usually equal with maturity; dry; bald or very finely hairy; whitish; with prominent, copious, white basal mycelium that often spreads through the substrate. **FLESH** white; odor and taste usually mealy, but sometimes not distinctive. **SPORE PRINT** white.

ECOLOGY: Likely mycorrhizal; fruiting on the litter of conifers, especially pine, spruce, and fir; growing scattered, gregariously, or in arcs or rings; summer and fall.

OBSERVATIONS: The gills of this mushroom are separable as a layer; try using your thumb to push them away from the cap. *Leucopaxillus albissimus* is one of the most common conifer associates in the Southern Rockies.

Photos by Vera S. Evenson

Neolentinus ponderosus (O. K. Miller) Redhead and Ginns

DESCRIPTION: CAP large, 10–35 cm across, convex to nearly plane, often with center depression; cinnamon to pinkish buff; dry surface, at maturity with low flattened scales. **GILLS** whitish to buff, distinctively serrated (saw-toothed) on edges, attached to slightly decurrent, close and narrow. **STALK** 4–10 cm long × 3–6 cm across; sometimes off-center, tough, narrowed at base; surface dry, finely scaly; top whitish buff, with scattered pale cinnamon scales below; basal area at maturity reddish brown and attached firmly to woody substrate; partial veil absent. **FLESH** tough, thick and white, not staining or decaying readily; odor mild and taste mild. **SPORES** buff in print.

ECOLOGY: A brown rotter of conifer wood, usually pine, this large mushroom is known only from western North America. Slow to decay, fruitings of this mushroom can be seen rotting stumps of dead ponderosa pines in midsummer.

OBSERVATIONS: The species epithet *ponderosus*, certainly is appropriate for this sometimes massive mushroom. Brown rotting fungi perform a valuable service in nature by producing very specialized enzymes that break down the celluloses in wood, leaving the brownish lignin-containing residues. The resulting brown cubical rot is crumbly and disintegrates relatively quickly back into the environment to replenish the soil. Not poisonous but needs extensive cooking because of its tough flesh.

Photo by Vera S. Evenson

Russula subalutacea Burlingham

DESCRIPTION: CAP fairly large, up to 10 cm across, convex, flattening with age; strong deep red to purplish red; margins even; surface glabrous, viscid; cuticle peels easily at least one-third to halfway. **GILLS** close, narrow, barely attached at stipe; whitish in young, becoming deep ocher from spore development at maturity. **STALK** 3–7 × 1.5–3 cm; thick and rather bulky, often wider at base; white with tinges of pale pink in some parts. **PARTIAL VEIL** absent. **FLESH** whitish, firm in youth, becoming rather fragile-brittle; not staining noticeably at injury; odor and taste mild. **SPORES** strong ocher to ocher-orange in print.

ECOLOGY: Usually two or three together; associated with pines in montane habitats, especially ponderosa pines, presumably mycorrhizal with these trees; fruiting in early to midsummer, erupting from under the pine duff after heavy rains.

OBSERVATIONS: Just one hundred years ago, in 1914, pioneering mycologist L. O. Overholts collected the type specimen in the Tolland area of Colorado "under pines." Since that time few additional collections of this very noticeable mushroom have been identified, probably because red Russulas are notoriously hard to identify as there are a lot of look-alikes. Another similar red-to-purplish, fairly large mushroom with a deep ocher spore print found under conifers is *Russula xerampelina* (Pg 96), commonly known as the "shrimp mushroom." Its odor is distinctly shrimpy and the cut flesh stains brownish.

Photo by Vera S. Evenson

Xeromphalina cauticinalis (Withering) Kühner and Maire

DESCRIPTION: CAP small, 1–2 cm across, broadly convex with slightly depressed center with age; margin slightly striate; orange-brown, centers red-brown; smooth, moist. **GILLS** subdistant, veined, decurrent; pale peach-colored, narrow. **STALK** long and narrow, 2–7 cm × 2–4 mm across, equal, tough, pliant; ocher-brown above and red-brown below, distinctively connected to substrate by orange-brown mycelium; partial veil absent. **FLESH** thin, pliant, pale ocher; odor and taste mild. **SPORES** white in print.

ECOLOGY: Gregarious to scattered on conifer debris and rotting pinecones; summer into fall. Even if they are small, these debris-rotting mushrooms and others like them have a combined impressive effect in conifer ecosystems by digesting and recycling woody debris that is hard to break down. Cones of all types are notoriously difficult to recycle.

OBSERVATIONS: A close relative, *Xeromphalina campanella* (Pg 175), is similar, but its growth habit is densely gregarious to caespitose on conifer wood, especially on decaying trunks. The caps of *X. campanella* are distinctly striate.

Photo by Vera S. Evenson

Boletus barrowsii Thiers and A. H. Smith

DESCRIPTION: CAP large, up to 25 cm across at maturity; convex, surface dry and fairly smooth, off-white to grayish buff. **TUBES** at first white, soon yellowish buff to olive from spore development; tube mouths (pores) olive-yellow at maturity; not staining at injury; tubes depressed where they meet the stalk. **STALK** bulky, 6–15 cm long × 3–7 cm thick; whitish to dingy brown; upper surface decorated with distinct netlike veins. **FLESH** firm, whitish, not turning blue where exposed or injured; odor mild and pleasant; taste pleasant to nutty. **SPORES** dark olive in print.

ECOLOGY: These distinctive boletes are mycorrhizal, gregarious to single in soil under conifers, usually pines; occurring in the Rocky Mountain region, especially in New Mexico and Colorado and also in California and the Pacific Northwest. They fruit in midsummer in warm exposed sites, often after rains. Commonly this large bolete fruits in the same ponderosa pine grove year after year.

OBSERVATIONS: The specific epithet honors Chuck Barrows, prolific collector from New Mexico, who often sent his western collections to Alexander Smith, professor at the University of Michigan. The more widely distributed *Boletus rubriceps* (Pg 179) is similar, but its cap colors are distinctively reddish brown and it is often viscid; *B. edulis*, the typical King Bolete is more brown. King boletes are mycorrhizal but have associations with many more types of native trees. All of these large *Boletus* are well-known and prized edibles.

Photo by Vera S. Evenson

Suillus kaibabensis Thiers

DESCRIPTION: CAP 4–12 cm across, convex, margins smooth to wavy or irregular with age; pale cinnamon to yellow-orange, often streaked; surface smooth and viscid-sticky, often becoming shiny upon drying. **TUBES** notched at stipe; whitish in young, becoming yellow-brown; tube mouths (pores) small, 1–2 per mm, angular, yellowish to brownish cinnamon with age, not staining noticeably at injury. **STALK** equal, 3–8 cm × 1–2 cm; whitish to yellow above, distinctly decorated with red-brown to tan glandular dots and tiny blobs; ring absent. **FLESH** soft, fairly thick, especially over the stalk; pale yellowish, not staining on bruising; odor and taste mild and pleasant. **SPORES** cinnamon-brown in print.

ECOLOGY: Mycorrhizal with pines, both ponderosa and lodgepole pines in the Rocky Mountains; usually gregarious, June to September. Common.

OBSERVATIONS: The true *Suillus granulatus* is a classic European species, and it is unclear whether North American versions represent the same phylogenetic species. In the Southern Rockies, the name *S. kaibabensis* has been applied to a pale look-alike that always grows under ponderosa pines; it is fairly commonly seen in the Four Corners area of the southwest. *Suillus brevipes* is similar with ocher-brown to brown sticky caps. It also has pine associations, especially lodgepole pine, but its stocky stalks are distinctly white and smooth, without glandular dots.

Photo by Vera S. Evenson

Auriscalpium vulgare Gray

DESCRIPTION: CAP small, 0.5–2.5 cm broad, convex to plane, eccentric, margins often finely fringed; dry; brownish, covered with dark brown fibrils. **SPINES** very fine, needlelike; 2–3 mm long; pale to medium brown; crowded. **STALK** slender, 3–7 cm × 2–3 mm; lateral; hairy; entire length clothed in dark brown fibrils; ring absent. **FLESH** firm, thin, white, and pliant; odor absent; taste unknown. **SPORES** white in print.

ECOLOGY: Single to gregarious, sometimes in clusters, attached to conifer cones, often ponderosa pinecones, typically on partly buried cone piles on the ground. These "spine fungi" produce their spores on sharp pointed spines located under the cap. They are wood recyclers that are widely distributed in the Rocky Mountains, summer and fall.

OBSERVATIONS: This interesting saprobe obtains its energy by producing specialized enzymes to digest the very resistant material in conifer cones. Without the action of these and other wood-rotting essential organisms in our forests, accumulations of massive piles of conifer debris would interfere with the health of the forest.

Photo by Ikuko Lubow

Fomitopsis cajanderi (Karsten) Kotlaba and Pouzar

DESCRIPTION: Perennial. **CAP** bracketlike, woody and tough, up to 20 cm across, shelving, hoof-shaped, joined at attachment to wood; upper surface distinctly tomentose and rosy pinkish to grayish pink, becoming brownish with age. **TUBES** short, grayish, tube mouths (pores) rose-colored, circular to angular. **STALK** absent, fruiting body growing directly on the substrate. **FLESH** tough to corky; pale pinkish brown; odor and taste mild. **SPORES** white in mass.

ECOLOGY: Growing on dead wood and stumps of conifers including ponderosa pine and Douglas fir in the montane zone. Visible throughout the year, these perennial wood-rotters are associated with a distinctive rotting action that turns the wood into brown cracked cubelike pieces that are eventually incorporated into native soils.

OBSERVATIONS: Recognized by its gorgeous rosy colors; *Fomitopsis rosea* is similar to *F. cajanderi* and also rots conifer wood, but it has different spores and other microscopic features. Brown rotting fungi have powerful enzymes that rot and recycle wood; they perform an essential service in forest environments by cleaning up dead wood and debris and recycling it back into the soil for the use of developing plant life.

Photo by Vera S. Evenson

Porodaedalea pini (Brotero) Murrill

DESCRIPTION: Perennial. **FRUITING BODY** hooflike, hard and woody, shelving, attached directly to tree bark; 5–20 cm broad; upper surface often with zones of appressed tomentum; brownish orange to deeper brown to blackish; cracked to grooved or ridged concentrically; tubes 5–6 mm deep; undersurface with brownish circular pores, 2–3 per mm, often labyrinthine to elongated. **STALK** absent. **FLESH** hard, woody; brownish orange. **SPORES** light to brownish with age.

ECOLOGY: Widely distributed in conifer forests, associated with a white rot of the heartwood of many conifers including ponderosa pines; on living or dead trunks; visible throughout the year. Common throughout North America.

OBSERVATIONS: Formerly called *Phellinus pini*, this powerful rotter causes a white pocket rot of our most economically important timber trees at a great loss to the timber industry. *Phellinus tremulae* is another similar wood-rotter, but it is found on living aspens in many of our Rocky Mountain areas.

Photo by Vera S. Evenson

Spathularia flavida Persoon

DESCRIPTION: FRUITING BODY consisting of a distinctive, compressed, fan-shaped head attached to a demarcated stalk; head up to 6 cm high × 1–3 cm across; flattened to irregularly lobed, distinctly decurrent and attached on opposite sides of the stalk; pale to medium yellow, often with orange tints; dry; smooth. **STALK** 2–3 cm long × 0.5–1.5 cm wide; fairly smooth, tapering toward base; yellowish with whitish mycelium at base; veil absent. **FLESH** thin; dry; odor and taste mild. **SPORES** whitish.

ECOLOGY: Gregarious to troops under conifers, pines, and firs in the Rocky Mountain region. This interesting member of the Ascomycota fruits in the late summer, saprobic in conifer litter and mosses; often called the yellow earth tongue.

OBSERVATIONS: *Neolecta vitellina* and its close relatives are similar-looking ascomycetes growing in conifer litter, but their fertile yellowish heads are not flattened and fan-shaped; nor are their heads markedly differentiated from their white bases—except by color. Members of the basidiomycete genus *Clavariadelphus* (Pg 182) are more club-shaped.

Photo by Cathy L. Cripps

DOUGLAS FIR FORESTS

Douglas fir (*Pseudotsuga menziesii*) forests in the Rockies are all about light and space. In this part of the country, the tall stately trees are well-spaced, which allows sunlight to penetrate to the forest floor. There is also enough space between trees for gentle breezes to ripple the treetops in waves and for avian inhabitants to maneuver. In the Pacific Northwest, giant Douglas fir trees (variety *menziesii*) form magnificent cathedral-like forests full of intriguing creatures such as northern spotted owls and flying squirrels. The Rocky Mountain Douglas fir is smaller (variety *glauca*), but it can still be impressive and these forests also have a rich ecology.

In the Northern Rockies, Douglas fir forms extensive pure forests that provide a transition zone between open landscapes and the denser higher elevation conifer forests. There are over 8 million acres of Douglas fir forests in Montana alone; pure forests are found on the east side of the Continental Divide along the Front Range, in the Bitterroots particularly along the Lochsa River and in the national forests of southwest Montana. The Lubrecht Experimental Forest near Missoula supports significant Douglas fir habitat. An ancient 500-year-old Douglas fir still survives in Wyoming near Yellowstone National Park. In the Southern Rockies, Douglas fir trees are more sporadic and are typically mixed with other conifers, but pure forests do occur. Douglas firs can be observed on the shaded side of valleys leading up into the mountains at lower elevations and mixed with other conifers on the Colorado Plateau, but purer forests exist in the Pike's Peak region. Ponderosa pines are more likely to occupy the montane zone at lower latitudes in the Southern Rockies.

Douglas fir is neither a fir (*Abies*) nor a spruce (*Picea*), and it is not a hemlock (*Tsuga*). The tree can be recognized by its thick corky bark with deep (sometimes orange) furrows. The bark and lower branches are often covered with the lime-green wolf lichen *Letharia vulpina*, a name resulting from the historic practice of using it to poison wolves in Europe. Douglas fir needles are flat like that of a fir, but arise singly and each has a small stalk, unlike true fir. The most definitive way to identify a Douglas fir is by its unique cones. Each cone scale has a papery bract tucked under it that looks like the back end of a mouse with two legs and a tail sticking out.

Because sunlight penetrates the canopy, the understory can be lush with grasses, herbs, and shrubs, or the forest floor can be almost barren with only a few needles sprinkled around where it is shaded. Plants such as snowberry, big huckleberry, Oregon grape, strawberry, service berry, gooseberry, buffaloberry, and bearberry attract a diversity of wildlife looking for their edible fruits. Silvery lupine, birchleaf spiraea, starry Solomon's seal, heart-leafed arnica, clematis, arrowleaf balsamroot, bedstraw, yarrow, and geraniums are some of the more conspicuous flowering plants of Rocky Mountain Douglas fir forests. Coralroot orchids (*Corallorhiza striata*) and pinedrops (*Pterospora andromedea*) are found in the shadier parts of the forest. Once considered parasitic plants because they lack green leaves, we now know these nongreen plants gain their nutrition by tapping into the connection between trees

Douglas fir has deeply furrowed bark, lower branches are often covered with lichens, and cones have mouselike bracts. *Photos by Cathy L. Cripps*

Silvery lupines under Douglas fir.
Photo by Cathy L. Cripps

Mycotrophs striped coralroot and pinedrops do not have green leaves and tap into mycorrhizal connections. *Photos by Cathy L. Cripps, Yellowstone National Park*

Day inhabitant of Douglas fir forests, the western tanager.
Photo by Andy Hogg

and mycorrhizal fungi. These "cheaters" are now called *mycotrophs* because sugars flow from the tree to the mycorrhizal fungus and into these plants in a three-way symbiosis.

Deer can be observed wandering through and hiding fawns in Douglas fir forests in spring, and bighorn sheep use these montane forests for cover and browsing in winter. A pine marten or weasel might be spotted resting on a branch. A variety of birds such as Cassin's finches, chickadees, nuthatches, crossbills, thrushes, western tanagers, evening grosbeaks, and sharp-shinned hawks enjoy the open forests and lush understories for feeding. Woodpeckers like to pick in the deep bark crevices for insects. Blue and ruffed grouse come for the berries and can surprise with a loud ruffle of wings when flushed from the understory.

However, like the magical old-growth Douglas fir forests of the Pacific Northwest, the real story might be said to occur at night when the forest's nocturnal inhabits awake—and mammals fly. Many bat species prefer open mature Douglas fir forests for maneuvering and for the availability of their favorite foods, such as moths. Long-eared bats such as *Myotis evotis* make these forests their home from Colorado to Canada. Douglas fir trees prefer calcareous soils and this means that karst topography and limestone caverns are sometimes nearby. Here, bats have the option of roosting in trees, on cliffs, in rock crevices, or in caves during the day. At night they locate prey by listening for the sound of insect wings on rough bark or using echolocation.

Another aerial mammal, the northern flying squirrel (*Glaucomys sabrinus*) also prefers open mature montane forests for night flying. The ecology of these squirrels is woven into the fabric of the forest, and they live in abandoned woodpecker cavities and use tree lichens as nesting material. Part of their diet consists of subterranean ascomycete and basidiomycete truffles, which they forage for at night by sniffing them out. The fungal spores are spread in their droppings, where they then germinate and reunite with their mycorrhizal partners, the Douglas fir trees.

Other nocturnal fliers such as flammulated owls (*Psiloscops flammeolus*) and great gray owls (*Strix nebulosa*) prefer montane Douglas fir and ponderosa pine forests. The open stands offer roosting and maneuvering room for the great grays with their four-foot wingspan. The tiny flammulated owls (less than 7 inches long) nest in tree cavities and go after moths at night like the bats. Interestingly, the "brooms" on large trees caused by the rust fungus *Chrysomyxa* or a parasitic mistletoe are used by martens, great gray owls, and flying squirrels as resting platforms or to support their nests. Trees are susceptible to Douglas-fir tussock moths and Douglas-fir bark beetles, but flying mammals and birds help keep insect populations in check.

It is best to look for fungi in Douglas fir forests in early spring and again in fall if the rains come, avoiding midsummer when forests can be dry. Mushrooms spring up early before higher elevation snow is melted out. The tiny *Strobilurus trullisatus*

Night flying mammals: Western long-eared bat and northern flying squirrel. *Photos by Amie Shovlain, Yellowstone National Park*

Great gray owl perched on the lichen *Letharia vulpina*; the large apothecial cups of *Letharia columbiana*. *Photos by Andy Hogg, Cathy L. Cripps*

and S. *albipilatus* are specialist decomposers of Douglas fir cones. The brilliant golden yellow polypore, *Laetiporus conifericola* can fruit on mature Douglas fir trees in fall in northwest Montana and Idaho. There are over 2,000 species of ectomycorrhizal fungi associated with Douglas fir, although most are known from the Pacific Northwest. The ectomycorrhizal fungi that primarily associate with Douglas fir include the orange-brown *Suillus lakei* and the lovely rose-colored but slimy *Gomphidius subroseus*. Shaggy Floccularias are also common, and some look similar to the yellow variety of *Amanita muscaria* found in Montana and Wyoming. In fall, the maroon to red-capped, fishy smelling *Russula xerampelina* can fruit in quantity.

Fire suppression policy has favored Douglas fir and it is now expanding into grasslands and regenerating in the understory of ponderosa pine forests. The moister forests with patches of open low ground cover are the best places to hunt fungi, because lush vegetation can obscure the mushrooms. However, berries might be ripe for the picking if fungi can't be found.

Photo by Cathy L. Cripps

Floccularia albolanaripes (G. F. Atkinson) Redhead (white form)

DESCRIPTION: CAP 5–9 cm across, convex; dry, smooth; white or cream, with a few pale yellow streaks or flattened scales in the center; margin turned under at first, with bits of yellow tissue on the edge. **GILLS** narrowly attached, fairly crowded; cream with slight pale yellow or pale pink tint. **STALK** 4–6 × 0.5–2 cm, club-shaped or equal; white and smooth above a ragged white ring that is tinted yellow; covered with floccose bits of tissue below. **FLESH** firm, white; odor not distinctive. **SPORE PRINT** white, weakly amyloid.

ECOLOGY: In small groups in Douglas fir forests in the Northern Rockies and in ponderosa pine and other conifer forests in Colorado; typically fruiting in summer in open areas in duff and needles. *Floccularia albolanaripes* is known only from the western United States and is relatively rare. Its ecology is not well defined.

OBSERVATIONS: *albolanaripes* for its white wooly stalk. *Floccularia albolanaripes* (G. F. Atkinson) Redhead has a bright yellow cap and this white form may be only a variant. Once called *Armillaria* because of their white spores and ring, *Floccularia* species are distinguished by their amyloid spores. Alex Smith and Sam Mitchel first described this form as *Armillaria albolanaripes* f. *alba* from Colorado. Flocculарias can be confused with *Amanita* and *Lepiota* species but do not have free gills. Not usually considered as an edible.

Photo by Cathy L. Cripps

Floccularia fusca (Mitchel and A. H. Smith) Bon

DESCRIPTION: CAP 4–7 cm across, convex, then flat and a bit wavy; dry or silky, covered with grayish brown flattened scales on a white background; sometimes cracking in the center. **GILLS** attached, dished near the stipe; white or cream, eroded with age; edges minutely scalloped. **STALK** 3–6 × 1–1.5 cm, narrowing slightly toward the base; white above a floccose ring; white with gray-brown scales below. **FLESH** firm, white; odor not distinctive or fungoid. **SPORE PRINT** white; spores amyloid.

ECOLOGY: Scattered in open needle duff in Douglas fir (Montana) or ponderosa pine and spruce-fir forests (Colorado and New Mexico); fruiting in summer and fall. This fungus is apparently known only from the western United States.

OBSERVATIONS: *fusca* for the grayish cap. *Floccularia* species are distinguished from each other by their cap and stipe colors; however, colors are variable making identification more difficult. *Floccularia fusca* lacks the yellow coloration found in other conifer associated species such as *F. albolanaripes* (Pg 93), *F. pitkinensis*, and *F. straminea* var. *americana* (Pg 107). Not considered an edible species.

Photo by Ed Barge

Lactarius rubrilacteus Hesler and A. H. Smith

DESCRIPTION: **CAP** 4.5–7.5 cm wide, convex with shallow depression in center, smooth, viscid; light brown, yellow-brown, light orange-brown, distinctly zoned, greenish tinged with age or when damaged; margin incurved at first. **GILLS** run slightly down the stalk, crowded; dull pink, pinkish gray, dull purplish red; greenish with age or when damaged. **MILK** deep purplish red; staining flesh purplish red, later greenish. **STALK** 3–4 × 2–2.5 cm; equal, straight or slightly curved; smooth, with a few pockmarks; dull pink (similar to gill color); bruising deep purplish red, eventually greenish; hollow when mature. **FLESH** thick; dingy white, quickly staining deep purplish red, then greenish when damaged; odor and taste mild. **SPORE PRINT** cream.

ECOLOGY: Mycorrhizal with conifers, reported with 2-needle and 3-needle pines, but most often with Douglas fir, fruiting in summer to fall; relatively rare in the Rockies and more common in the Pacific Northwest.

OBSERVATIONS: *rubrilacteus* for the red milk. The orange cap and red milk are reminiscent of *L. sanguifluus*, which is now considered a European species. The flesh of the more well-known edible *L. deliciosus* (Pg 169) also turns green but its latex is carrot-colored. *Lactarius rubrilacteus* is eaten at least in the Pacific Northwest and in Colorado.

Photo by Michael Kuo

Russula cf. xerampelina (Schaeffer) Fries

DESCRIPTION: CAP 2–9 cm across; convex, becoming broadly convex to flat; dry; bald and shiny; bright yellow when first emerging, soon turning maroon, red, dull to bright brownish orange, or orangish brown, sometimes mottled with purplish, greenish, tan, yellowish, or reddish areas; the margin not lined; the skin peeling roughly one-third to three-fourths of the distance to the center. **GILLS** attached to the stem; close or nearly distant; white, maturing to pale yellow, sometimes discoloring brown. **STALK** 3–7 cm long; up to 3.5 cm thick; more or less equal; bald; whitish (sometimes with a pink tint), slowly bruising brown where handled (allow up to one hour). **FLESH** white; brittle; not changing when sliced; odor faint when young, later fishy or shrimplike and taste mild. **SPORE PRINT** yellow, spores with isolated amyloid warts.

ECOLOGY: Mycorrhizal with conifers, especially Douglas fir in the northern states; growing alone, scattered, or gregariously; late summer and fall; widely distributed in the Rocky Mountains.

OBSERVATIONS: This is the Rocky Mountain version of *Russula xerampelina*, a traditional European species with many North American representatives; it is likely that our version is a distinct, unnamed species (or group of species, given the variability in cap colors). A drop of iron salts on the stems of species in this group turns green. The species tends to be more maroon in the north and redder in the Southern Rockies. Considered edible, but there are numerous look-alikes, and some are toxic. The slow brown bruising reaction can be more reliable than the fishy smell; species that are hot when tasted raw should be avoided.

Photo by Vera S. Evenson

Strobilurus albipilatus (Peck) V. L. Wells and Kempton

DESCRIPTION: CAP 1–1.5 cm across; shallow convex, or a bit conic-convex, smooth; brown, pale brown, gray-brown, yellow-brown, paling; margin slightly lined. **GILLS** attached, relatively well-separated; white, cream, or pale yellow-brown. **STIPE** long and thin, 1.5–4 × 0.1–0.2 cm, equal; yellowish golden, paler at the apex; smooth; often with white mycelium at the base. **FLESH** pale; odor not distinctive. **SPORE PRINT** white, not amyloid.

ECOLOGY: A saprobe of old cones or conifer debris in Douglas fir, spruce-fir, or mixed conifer forests. Typically fruiting in early spring and early summer, sometimes near melting snowbanks at higher elevations; it can be one of the first fungi to fruit in Douglas fir forests at lower elevations; occurring in the Rocky Mountains and Pacific Northwest.

OBSERVATIONS: *albipilatus* for its light cap color because it pales on weathering or drying. It is similar to *S. trullisatus*, which is whiter, more restricted to Douglas fir cones, and occurs later in the year. These fungi are fascinating because of their rather strict association with cones and conifer debris. Often excavation is necessary to find the substrate that helps in the identification.

Photo by Cathy L. Cripps

Gomphidius subroseus Kauffman

DESCRIPTION: CAP 2–5 cm across; convex becoming flat and dished in the center, funnel-like with age; very slimy; rosy pink to rosy red when young, becoming dull pink with age; margin rolled under at first. **GILLS** run down the stalk, thick, well-separated; white at first and covered with a slimy veil, becoming smoky gray or blackish at maturity. **STALK** 4–8 × 1–2 cm, tapering toward the base, smooth, sticky; white, bright yellow at the base; with a thin veil line often blackened with spores. **FLESH** white, bright yellow at the base; odor not distinctive. **SPORE PRINT** smoky black.

ECOLOGY: Thought to be mycorrhizal with Douglas fir in montane forests and occasionally with other conifers throughout the Rockies and the West, typically fruiting in scattered groups in summer. It is often found fruiting with *Suillus lakei* and there is some evidence of a physiological connection.

OBSERVATIONS: *subroseus* for its color. The gills are white at first so the black spore print is a surprise. This small species is recognized by its slimy pink cap, funnel-shape, yellow at the base of the stalk, and association with Douglas fir. *Gomphidius glutinosus* is also yellow at the base, but is larger, has a brown cap, and is often with other conifers. Both are considered edible, but the slime layer definitely needs to be removed.

Photo by Cathy L. Cripps

Suillus lakei (Murrill) A. H. Smith and Thiers

DESCRIPTION: CAP 4–10 cm across or larger, convex, sticky or dry; dark to dull reddish brown or orange-brown, with flattened scales; margin rimmed with white tissue or not. **PORES** small and shallow, becoming larger and radiating in rows; bright lemon-yellow to golden, staining red-brown. **STALK** 4–6(–12) × 1–3 cm, slightly swollen at the base; yellow above the superior ring, which is a band of whitish to pale yellow tissue; dingy pale orange below. **FLESH** solid, bright lemon-yellow, staining bluish green at the base of the stipe when cut or bruised; odor not distinctive. **SPORE PRINT** cinnamon-brown.

ECOLOGY: Mycorrhizal mostly with Douglas fir but also found in mixed forest types in montane Rocky Mountain and Pacific Northwest forests; one of the first *Suillus* species to fruit in early summer and common in wet weather.

OBSERVATIONS: *lakei* for Oregon professor E. R. Lake. A number of varieties have been described, but here we consider it as one species for the Rockies. The cap surface varies from dry fibrillose with flat scales to smooth and slimy, a condition that is weather-dependent. It could be confused with *S. tomentosus*, which has a dry scaly cap, but that species is dull gold, occurs with lodgepole pine, and has flesh that turns bluish when cut open. *Suillus lakei* is a meaty mushroom and is considered edible but not tasty by some.

Photos by Cathy L. Cripps

Cryptoporus volvatus (Peck) Shear

DESCRIPTION: FRUITING BODY up to 5 × 5 cm across and high; knob-shaped with a rounded top that is smooth, dry, hard; cream; attached on one side to trees; margin joined seamlessly to the lower covering. **PORES** tiny; cream; hidden by a thick tough membrane that often has a small hole or two from insect activity; tubes rather thick when cut open. **FLESH** tough, corky; odor slightly resinous or fragrant. **SPORES** whitish.

ECOLOGY: Scattered on standing or recently killed conifers that still have their bark; found on Douglas fir, as well as pines, grand fir (in the north), and occasionally other conifers; fruiting throughout the season; a weak white rot of sapwood. Widespread in western North America and in northern states; not known in Europe.

OBSERVATIONS: *volvatus* for the membrane and *crypto* because the pores are hidden. This polypore is often observed on trees that have been beetle-killed, and the beetles may move the spores from tree to tree giving the fungus a head start; it is eventually replaced by other decomposer fungi.

Photo by Cathy L. Cripps

ASPEN FORESTS

The iconic quaking aspen (*Populus tremuloides*) forests of the Rocky Mountains stretch from low elevations up into the subalpine, punctuating extensive portions of the landscape. The aspen's white bark and bright green leaves stand in stark contrast to the dark green conifer forests of the West. In spring, the heart-shaped leaves are a luminescent lime-green, which deepens to emerald in summer and senesces to brilliant shades of yellow and orange in fall. These are the golden forests tourists come to see blazing in the late autumn light. Aspen leaves tremble and rustle in the wind, which together with their color make them one of the most recognizable trees by eye or ear in the Rockies. The extensive pure aspen forests of Colorado and Utah give way to isolated stands that merely dot the landscape in the Middle Rockies, but their grandeur is regained in the vast Aspen Parklands near the Canadian border in Alberta.

One aspen stand in Utah has been named as one of the largest, heaviest, and oldest organisms on earth. It is the clonal nature of aspens that allows this designation

Aspen's heart-shaped leaves with a twisted petiole. *Photo by Cathy L. Cripps*

In autumn aspen leaves turn a blazing orange-yellow. *Photo by Carol Schmudde*

since whole forests are actually composed of sprouts (ramets) that arise from one giant root system. The Utah aspen clone called *Pando* covers 106 acres, is calculated to weigh 6,600 tons, and is estimated to be 80,000 years old. Walking through an old aspen clone forest gives a deep sense of time and place; some forests may predate Rocky Mountain glaciers. Aspen stands can also be transitory (seral), giving way to the invasion of conifers, but where it persists, aspen defines organism longevity. Forest fires help reset the successional clock and allow light and space for new aspen sprouts. Sprouts can be observed by the hundreds springing from old root systems after wildfires. While aspen also produce tiny seeds in long catkins, it is rare that a seedling actually establishes except at the northern limit of its range. However, aspen seedlings were reported after the great 1988 forest fires in Yellowstone National Park.

Elk (wapiti) are browsers of aspen leaves and sprouts. *Photo by Cathy L. Cripps*

Aspen forests are home to an amazing array of plants, animals, and birds, as well as mushrooms. White-tailed and mule deer (*Odocoileus*) and elk (*Cervus canadensis*) are prolific browsers of upland aspen. In areas where these large mammals are numerous, they are suspected of preventing aspen regeneration because they can consume a voluminous number of sprouts. Aspen bark is also a source of food for elk in winter. In areas where streams intersect aspen stands, beavers (*Castor canadensis*) feed on the bark and small branches and gnaw down trees to incorporate in their dams. One beaver can eat 1–2 pounds

Red-naped sapsucker at its cavity nest and an aspen conk to the right of the hole. Birds prefer trees softened by this decay fungus, *Phellinus tremulae. Photos by Cathy L. Cripps*

of aspen a day and when multiplied by countless animals, a monumental amount is consumed by beavers over the course of a summer. Voles and other small mammals gnaw at the lower bark under snow in winter, giving the base of older aspen trees a characteristic furrowed gray crusty appearance.

Birds are numerous in aspen forests, and many cavity-nesting birds make their nests in the trunks of old aspen. The red-naped sapsucker (*Sphyrapicus nuchalis*), a common inhabitant of aspen forests, often chooses to excavate its nest in dead and dying aspens previously softened by the aspen conk fungus (*Phellinus tremulae*). Chirping young sapsuckers can give away the location of nests in spring. Bears have been known to climb aspens to locate bird nests, for cub safety, and also to eat catkins or buds in spring. Claw marks left in the tender aspen bark indicate their previous activity. Similarly, carvings in aspen bark by Mexican and Basque sheepherders provide historical evidence of a bygone era.

Understory plants are as varied as the soils, topography, elevation, and climate where aspen can exist and include snowberry, chokecherry, corn lily, bracken ferns, and an array of wild flowers such as the black-eyed Susan, wild rose, and chiming bells. The state flower of Colorado, the strikingly beautiful blue columbine (*Aquilegia caerulea*) can be observed at the edge of aspen forests in this state but it is the yellow columbine (*A. flavescens*) that festoons adjacent meadows in Northern Rocky Mountain forests.

Huge aspen forests are spread over the Unita and Wasatch mountains in Utah, the San Juan and Elk Mountains of Colorado, areas east of Glacier National Park, and the Aspen Parklands of Alberta. Some of the largest trees are found along Kebler Pass in western Colorado and pygmy forests of stunted aspen occur at the U.S.–Canadian border near Glacier and Waterton National Parks. Early successional aspen can be observed around many defunct metal smelters (Butte/Anaconda, Montana) where toxic air from processing wiped out the previous vegetation. Sudden aspen decline (SAD) due to many stress-related factors has reduced stand health in many areas, particularly in the Southern Rockies.

Blue columbines of the Southern Rockies and yellow columbines in Montana. *Photos by Lee Gillman, Cathy L. Cripps*

Since aspen exists on a variety of soil types at various elevations, it pays to select the right forest type during the right season for collecting mushrooms. Spring edibles such as black morels and aspen oysters and fall prizes such as wild enoki (*Flammulina populicola*) are found in season. July is often an interesting time to collect under aspen as this is when orange-capped boletes (*Leccinum* species) can abound. Mycorrhizal mushrooms such as *Paxillus vernalis*, *Lactarius controversus*, and many *Leccinum* species are unique to aspen forests and are found nowhere else. Tiny decomposers including *Tubaria*, *Crepidotus*, *Simocybe*, *Flammulaster*, *Heliocybe sulcata*, and *Peniophora rufa* prefer aspen wood and twigs. Walking through one of the world's largest organisms on a sunny summer day is luscious enough, but the tremendous biodiversity it attracts makes aspen forests a worthwhile place to collect mushrooms.

Mycorrhizal orange-capped boletes nestled in the grass of an aspen stand. *Photo by Cathy L. Cripps*

Aspen oysters decomposing aspen bark. *Photo by Michael Kuo*

Photo by Vera S. Evenson

Amanita 'stannea' nom. herb.

DESCRIPTION: CAP 4–9 cm across, convex to plane with slight umbo; striated one-third the distance to the center; surface viscid in young; distinctively light tin-gray, covered with scattered white patchy remnants of the volva; patches becoming orange-brown at their edges. **GILLS** free, white to tinted fawn-colored, close; margins white at first, soon developing drab brownish edges. **STALK** 3–8 × 1–2 cm; surface white, midsurface with grayish fibers, often in a herringbone pattern; equal, not bulbous; ring absent; universal veil white, rather thick, friable, basal part remaining as a cup or breaking up, stained orange-brown here and there like the patches on the cap. **FLESH** white, thin, not staining; odor mild, do not taste. **SPORE PRINT** white.

ECOLOGY: Gregarious, growing in soil near aspens; presumed to be mycorrhizal with aspens; midsummer to early fall. Known from aspen-forested areas of Colorado and New Mexico in the Southern Rocky Mountains, and not uncommon. Likely the same species as that recorded as *A. constricta* from Montana aspen stands.

OBSERVATIONS: This lovely exannulate *Amanita* has been recognized at least since Dr. Alexander Smith and other mycologists collected it at the Aspen Mushroom Conference in the late 1970s. It was given the provisional name *A. stannea* there, which refers to the color of tin. Two other taxa, *A. protecta* and *A. constricta* are similar and one also has a volva that stains orange, but they are from oak forests in California. There is no DNA evidence available to connect them, so we elect to use Smith's herbarium name here.

Photo by Michael Kuo

Flammulina populicola Redhead and R. H. Petersen

DESCRIPTION: CAP 1–7 cm across; convex, becoming broadly convex to flat; moist and sticky when fresh; bald; orangish brown to yellowish brown. **GILLS** attached to the stalk; whitish to pale yellow; crowded or close. **STALK** 2–11 cm long; 3–5 mm thick; equal, or larger toward base; tough; pale to yellowish brown or orange-brown when young, becoming covered with a dark brown to rusty brown or blackish velvety coating as it matures. **FLESH** whitish to yellowish, thin; odor and taste not distinctive. **SPORE PRINT** white.

ECOLOGY: Saprobic on the stumps, logs, roots, and living wood of aspens and cottonwoods; growing alone, scattered, gregariously, or (more commonly) in clusters; often appearing terrestrial when growing from root systems; late summer and fall, often one of the last mushrooms to fruit.

OBSERVATIONS: *populicola* for its host. This species is similar to the well-known *Flammulina velutipes* but differs in habitat and microscopic features. *Flammulina velutipes* is commonly cultivated and is called enoki or enokitake. Both species fit the name of "velvet stems."

Photo by Melissa Kuo

Floccularia straminea var. americana (Mitchel and A. H. Smith) Bon

DESCRIPTION: CAP 4–18 cm across; convex or nearly round when young, becoming planoconvex, flat, or broadly bell-shaped; dry; conspicuously scaly with soft, vaguely concentrically arranged scales, especially in the outer half; bright yellow when young, but often fading to straw yellow or pale yellow; the edge adorned with partial veil remnants. **GILLS** attached to the stalk, close; yellow to pale yellow. **STALK** 4–12 cm long, up to 2.5 cm thick; more or less equal; white and smooth near the apex; sheathed below with shaggy zones of soft yellow scales in roughly concentric zones, sometimes with a flimsy ring. **FLESH** white, unchanging on exposure; odor and taste not distinctive. **SPORE PRINT** white.

ECOLOGY: Possibly saprobic in contrast to other Floccularias but the ecology is not confirmed; growing alone, scattered, or gregariously in quaking aspen stands and in spruce-fir forests; summer and fall.

OBSERVATIONS: *straminea* for its straw color. When fresh, this striking mushroom stands out in the understory. It is also sometimes given the European name *Floccularia luteovirens*. It can be confused with the yellow form of *Amanita muscaria* (Pg 165) in the Northern Rockies, which also occurs under aspen and conifers.

Photo by Ed Barge

Heliocybe sulcata (Berkeley) Redhead and Ginns

DESCRIPTION: CAP 1.5–4 cm, convex, hemispheric; cinnamon-brown or orange-brown, covered with small cinnamon-colored scales that can be darker in the center; margin turned down and strongly pleated. **GILLS** attached, rather thick, with saw-toothed edges; cream. **STALK** 1–3 × 0.4–0.6 cm, equal; pale cream at top often with pink tints, lower part covered with fine cinnamon-colored scales. **FLESH** whitish, tough and rubbery, especially in stalk; odor faint. **SPORE PRINT** white.

ECOLOGY: A saprobe almost exclusively restricted to downed aspen that has lost its bark; fruiting throughout the summer and into fall; a brown-rot fungus that preferentially decomposes cellulose.

OBSERVATIONS: *sulcata* for its pleated margin. This species is recognized by its scaly cap, toothed gill edges, and tough flesh. Once called *Lentinus sulcatus*, DNA analysis places it closer to the polypores than to gilled mushrooms. Its closest relatives, *Gloeophyllum* (Pg 139) and *Neolentinus* species (Pg 79) also have tough flesh. It might be confused with *Flammulaster*, *Crepidotus*, and *Simocybe* species, which are common on aspen wood without its bark, but these produce fragile mushrooms.

Photo by Michael Kuo

Lactarius controversus Persoon

DESCRIPTION: CAP 7–30 cm across; at first convex with an inrolled, slightly hairy margin; becoming flat with a central depression, or vase-shaped, with an even and bald margin at maturity; slimy to sticky when fresh, but soon dry; with appressed fibers; whitish overall, but often with faint zones of pinkish or purplish, especially near the margin. **GILLS** attached to the stalk or beginning to run down it, thin, close or nearly crowded, sometimes forking near the stem, distinctly pinkish to pale pink. **STALK** 2.5–10 cm long; 1.5–4 cm thick; more or less equal, or tapering to the base; sticky when fresh, but soon dry, usually without potholes, but occasionally with a few, eventually becoming hollow; whitish. **MILK** white, unchanging on exposure to air, not staining tissues. **FLESH** white, unchanging on exposure, fairly firm; odor not distinctive, or pleasantly fragrant; taste slowly moderately to strongly acrid. **SPORE PRINT** creamy white or pale pinkish; spores with amyloid ridges.

ECOLOGY: Mycorrhizal with aspen, and occasionally with willows; growing alone or gregariously, sometimes buried in leaf litter; summer and fall. Also known with birch in Europe.

OBSERVATIONS: The pink mature gills and the association with aspen make this inedible species easy to identify. Other whitish *Lactarius* species with aspen typically have a hairy margin and/or a more zonate cap.

Photo by Michael Kuo

Pleurotus populinus O. Hilber and O. K. Miller

DESCRIPTION: CAP 6–8 cm across (or larger); almost flat, shelf-shaped; greasy, smooth, slightly translucent (pearl luster); cream or buff; with a few lines at the margin, which is slightly turned under. **GILLS** white, cream, radiating from one underside of the cap and running down the stalk (if there is one). **STALK** absent or very short, 1 cm × 1 cm, white to cream, attached to one side of the cap. **FLESH** watery white; rubbery tough in the stalk; odor of anise or licorice. **SPORE PRINT** whitish.

ECOLOGY: Occurring as layers of shelving caps on standing or downed aspen wood (more rarely, cottonwood); fruiting in spring and early summer, often one of the first edible fungi to appear; a saprobic white-rot fungus that decays aspen wood to a white stringy texture. The species appears restricted to North America.

OBSERVATIONS: *populinus* for its occurrence on aspen. The other oyster mushroom of the Rocky Mountains, *P. pulmonarius* (Pg 57), has a darker gray-brown cap with a lilac or gray spore print and is typically on cottonwoods. Both species are excellent edibles but the flesh is compromised if insects find them first. If they are carefully harvested so as not to disturb the mycelium in the wood, they may continue to fruit.

Photo by Michael Kuo

Russula aeruginea Lindblad

DESCRIPTION: CAP 5–9 cm across; convex when young, becoming broadly convex to flat with a shallow depression; dry or slightly moist; smooth, or minutely velvety over the center; grayish green to yellowish green; the margin often lined by maturity; the skin peeling about halfway to the center. **GILLS** attached to or running slightly down the stalk; close, often forking near the stem; cream to pale yellow, sometimes becoming spotted brownish in places. **STALK** 4–6 cm long; 1–2 cm thick; whitish; dry; smooth; discoloring brownish in places, especially near the base. **FLESH** white; brittle; unchanging on exposure; odor and taste not distinctive. **SPORE PRINT** cream to pale yellow, spores with amyloid ridges.

ECOLOGY: Mycorrhizal with hardwoods or conifers, commonly found under aspen in the Rocky Mountains; growing alone, scattered, or gregariously; summer and fall.

OBSERVATIONS: *aeruginea* for its green color. The evenly green cap and mild taste distinguish this mushroom from similar species in the Rocky Mountains. Russulas in the *cyanoxantha* group are also found with aspen; they are typically mottled with pinkish purplish colors, but can appear greenish when the other color tones are faint.

Photo by Cathy L. Cripps

Tubaria furfuracea (Persoon) Gillet

DESCRIPTION: CAP 1–3 cm in diameter, convex, broadly convex or almost flat, smooth, a bit greasy; pale to medium orange-brown, drying lighter; margin with faint lines and sometimes hung with bits of white tissue. **GILLS** broadly attached or running slightly down the stalk; pale orange-brown. **STALK** 2–4 cm × 3–4 mm, long and thin, equal or slightly club-shaped; smooth; cream to pale orange; sometimes with a faint ragged ring zone at top. **FLESH** watery pale orange; odor not distinctive. **SPORE PRINT** pale brown.

ECOLOGY: This saprobe occurs in aspen or cottonwood stands on debris such as matted leaves, tiny twigs, or dead wood, often fruiting in large troops; fruiting in early spring and even recorded during warm weather in March and April in Montana and Idaho in places where the snow has melted and the soil has warmed. Also reported from Colorado.

OBSERVATIONS: *furfuracea* for the white fibrils on the cap margin. This "litter decomposer" is one of the first mushrooms to fruit in spring and the habitat and fruiting time help identify it. The overall orange-brown color and small bits of white tissue on the margin of young caps are also useful identification characters. Spring look-alikes in aspen forests include the more delicate *Flammulaster* species, which are covered with branlike flakes, have brownish spore prints, and also fruit later in the year.

Photo by Michael Kuo

Entoloma lividoalbum (Kühner and Romagnesi) Kubička

DESCRIPTION: CAP 5–7 cm across; conico-convex to bell-shaped or convex at first, becoming broadly convex, broadly bell-shaped, or nearly flat; greasy when fresh; bald; yellow-brown, fading with age; the margin not lined or only faintly lined at maturity. **GILLS** narrowly attached to the stem; close or nearly distant; at first white, becoming pink with maturity. **STALK** 5–8 cm long; 1–2 cm thick; more or less equal; dry; bald but finely lined longitudinally; white, often discoloring and bruising brownish near the base. **FLESH** white; thin; odor and taste mealy. **SPORE PRINT** pink.

ECOLOGY: Growing alone, scattered, or gregariously under hardwoods including aspen, but occasionally reported under conifers; summer and fall; large Entolomas in this group are suspected of being mycorrhizal but this is not confirmed.

OBSERVATIONS: This potentially poisonous species is somewhat reminiscent of a *Tricholoma*, but features a pink spore print (usually pinkish brown). The genus can be confirmed microscopically by the angular spores, but there are numerous species and they are difficult to distinguish.

Photos by Michael Kuo

Cortinarius trivialis J. E. Lange

DESCRIPTION: CAP 4–7 cm across, convex, orange-brown, smooth, covered with a thick slime layer especially at the margin. **GILLS** attached pale whitish or grayish, becoming pale orange-brown, covered with a gelatinous veil when young. **STALK** 4–9 × 1–2.5 cm, sometimes with a small bulb at the base; white to pale watery brown, more so at the base; covered with bands of slime that break up into scales. **FLESH** white to cream, brown and tough in stalk base; odor not distinctive. **SPORE PRINT** rusty brown.

ECOLOGY: Mycorrhizal with aspen in the Rocky Mountains, fruiting from early summer to fall; often found scattered in grassy areas, sometimes in great numbers. It is also known from Europe with *Populus tremula* and possibly other deciduous trees.

OBSERVATIONS: *trivialis* because it is "common," at least under aspen. Easily recognized by the slimy orange cap and white stalk covered with bands of slime when found under aspen. Its counterparts in alpine habitats are *C. favrei* (Pg 234) and the larger *C. absarokensis*, which occur with willows. Other *Cortinarius* species with both a slimy cap and slimy banded stalk (subgenus *Myxacium*) are typically mycorrhizal with conifers in the Rocky Mountains. None is edible.

Photo by Michael Kuo

Hebeloma insigne A. H. Smith, V. S. Evenson, and D. H. Mitchel

DESCRIPTION: CAP 5–10 cm across; convex, becoming broadly convex or nearly flat; sticky when fresh; bald; with a soft, cottony margin when young; cinnamon-brown to pinkish brown; the young margin adorned with fibers from the ephemeral cortina. **GILLS** attached to the stalk, often by a notch; close, moderately broad; pale clay color when young, becoming cinnamon-brown to brown; sometimes featuring beads of liquid when young and fresh; the edges often becoming ragged with age. **STALK** 4–8 cm long; 1–3 cm thick; more or less equal above a swollen base; whitish powdery near top, developing whitish scales below, often in more or less concentric bands; dry; the scales often capture spores as the mushroom matures and thus become brownish. **FLESH** thick; whitish; odor pungent, or not distinctive and taste not distinctive. **SPORE PRINT** dull brown to reddish brown.

ECOLOGY: Mycorrhizal with aspen, but often found in mixed aspen-conifer forests; usually growing gregariously; late summer and early fall. Reported from the Southern and Northern Rocky Mountains.

OBSERVATIONS: This large, aspen-associated species is similar to the well-known *Hebeloma sinapizans*, but features a cortina when very young and lacks the radishlike odor of that species. It should be regarded as potentially poisonous.

Photo by Michael Kuo

Paxillus vernalis Watling

DESCRIPTION: CAP 6–15 cm across, sometimes larger; shallow convex and dished in the center; smooth, tacky or with the feel of kidskin when dry; white to cream when young, becoming mottled yellow-brown and red-brown with age or on handling, eventually all brown; margin usually ribbed and rolled under. **GILLS** run down the stalk, some forking; easily separated from the flesh; white to cream, becoming red-brown with age or on bruising. **STALK** short and thick, 3–6 × 2–3 cm, tapering toward the base; dingy cream, staining red-brown. **FLESH** yellow, bruising red-brown; very firm; odor somewhat unpleasant. **SPORE PRINT** red-brown.

ECOLOGY: Mycorrhizal mostly with aspen but also with cottonwoods; fruiting in groups or in large numbers in spring, summer, and fall during rainy periods. Stalks are buried in the ground and caps can be partially buried in leaf litter; fruiting bodies are remarkably persistent.

OBSERVATIONS: *vernalis* for its appearance in spring, but it also fruits in the fall if conditions are wet enough. The whitish fruiting bodies when young and the dark red-brown spore print help distinguish it from *Paxillus involutus*, which is always brown and has yellow-brown spores. Both *Paxillus* species have been reported to cause gastrointestinal upset and may cause more serious poisoning (hemolytic anemia).

Photo by Michael Kuo

Pholiota squarrosa (Vahl) P. Kummer

DESCRIPTION: CAP 4–9 cm across, convex; pale yellow-brown; covered with flat or recurved fibrous brown scales over most of the surface; margin often with veil tissue. **GILLS** narrowly attached, crowded, narrow; pale yellow, pale brown to rusty brown, sometimes with a greenish tinge. **STALK** 6–12 × 1.0–2.5 cm, stout, equal or with small bulb; cream and smooth above the ragged ring zone, covered with yellow-brown scales below. **FLESH** pale yellow to brown, very firm in stalk; odor faint, or faintly of garlic or onion. **SPORE PRINT** brown.

ECOLOGY: In dense clusters springing from the base of aspens in summer, fruiting bodies can persist for weeks. It is rather widely distributed and also known from conifer forests in the Rockies.

OBSERVATIONS: *squarrosa* for its scaly cap. This species is recognized by its dense clusters at the base of trees and scaly caps. *Flammulina populicola* (Pg 106) and *Armillaria* species also grow in clusters at the base of aspens, but neither has scaly caps and stalks. This *Pholiota* has been listed as edible in earlier field guides, but there is evidence it can produce gastric upset, so it is not recommended.

Photo by Cathy L. Cripps

Psathyrella uliginicola McKnight and A. H. Smith

DESCRIPTION: CAP 4–8 cm across, convex and slightly domed, opening to almost flat; a bit greasy, smooth, or covered with fine blackish hairs; pale gray, gray-brown; margin pleated and turned under at first, later wavy and uplifted. **GILLS** close, with a narrow attachment; rather broad; white then gray with a slight lavender tint. **STALK** 4–8 × 0.8–1.5 cm, sometimes slightly larger at the base or apex; white; covered with minute hairs, hollow. **FLESH** white, firm but brittle; odor slight. **SPORES** blackish brown.

ECOLOGY: Typically on wet soil near aspen, fruiting in summer and fall; a characteristic species of Rocky Mountain aspen forests, but also with cottonwoods.

OBSERVATIONS: despite the name *uliginicola*, which suggests a wood-loving species, this *Psathyrella* usually fruits on wet soil, although aspen is usually nearby. The robust cap and the pleated margin when young along with the association with aspen help identify it. The gray cap can appear smooth, but McKnight mentions minute fibrils on the cap and stem in his original description. Not edible.

Photo by Cathy L. Cripps

Leccinum insigne complex Smith, Thiers, and Watling

DESCRIPTION: CAP 6–15 cm across, hemispherical, then broadly convex; slightly textured at first; dark red-brown becoming bright orange-brown; cuticle overhangs the cap margin a few mm. **TUBES** indented at stalk; pores tiny and cream, bruising brownish. **STALK** 10–15 × 4–6 cm, clavate, with white base color; covered with fine *scabers* (clustered hairs) that are pale brown and darken to blackish brown giving the stalk a rough texture; often bluish at the base. **FLESH** white, changing to wine pink (blush) when cut, then grayish; odor not distinctive. **SPORE PRINT** brown.

ECOLOGY: This mycorrhizal genus is very common in aspen forests in the Rocky Mountains; fruiting in summer from late June through July into August. *Leccinum* species can be dominant under aspen, especially in grassy areas; species in this complex are also found with aspen outside the Rockies.

OBSERVATIONS: There are a number of *Leccinum* species that occur under aspen and the names have not yet been resolved. The *Leccinum insigne-aurantiacum* group of orange-capped boletes needs to be sorted out in western North America; *L. aurantiacum* is a European name that probably does not apply. There are reports of serious gastric upset with *Leccinum* species especially when undercooked or eaten raw. We do not recommend eating *Leccinum* species since the basis of the toxicity is not yet known.

Photo by Cathy L. Cripps

Artomyces pyxidatus (Persoon) Jülich

DESCRIPTION: FRUITING BODY 4–8 cm wide × 6–8 cm high, coral-like, with numerous upright branches radiating from a base; those branches divide again and terminate in a crown of fine tips; surface smooth; pale yellow or yellow-brown, sometimes with a pinkish tinge, darker on tips with age. **FLESH** pale whitish; somewhat tough; odor strongly aromatic or faint. **SPORES** light-colored, amyloid.

ECOLOGY: A saprobe on downed aspen wood that is devoid of bark; usually in moist areas where streams run through aspen forests; generally fruiting in late spring and early summer, late June to July. Not limited to the West; also known from the Midwest and West coast where aspen occurs.

OBSERVATION: *pyxidatus* for the small boxlike structure of the tips; once called the more apt name *Clavicorona pyxidata* for the crownlike branch ends. It is one of the few coralloid fungi to fruit on wood, typically aspen. It could be confused with the slender coral *Ramaria stricta*, but the latter is on conifer wood and lacks "crowns." *Artomyces* species are not related to other coral fungi even though they have the same morpho-form. Not considered an edible species.

Photo by Cathy L. Cripps

Trichaptum biforme (Fries) Ryvarden

DESCRIPTION: FRUITING BODY up to 6 cm across, but larger when growing in compound groups; shelflike or petaloid, attached to wood on one side; upper surface fuzzy becoming smooth with age, with faint zones especially toward the margin; grayish with lavender tints, but fading; margin sharp and often lavender, especially on lower edge. **PORES** small and angular, becoming toothlike with age; pale to deep rich lavender, paling with age to buff; tubes shallow, 2–5 mm deep. **FLESH** tough, fibrous, pale buff; odor not distinctive. **SPORES** white.

ECOLOGY: A very common saprobe that is a white rotter of hardwoods throughout North America; in the Rockies it is found mostly on aspen, fruiting in summer and fall.

OBSERVATIONS: This polypore is recognized by its strikingly lavender toothlike pores when fresh, but colors fade with age to pale whitish buff, making it more difficult to identify. *Trichaptum subchartaceum* also has a lavender pore surface and is found on cottonwoods and aspen; it is larger, with thicker flesh, and the pores do not become as toothlike. We have seen both in the Rockies.

Photo by Don Bachman

LODGEPOLE PINE FORESTS

Lodgepole pine forests are dense and dark, filled with soldierlike trees standing at attention with rigidly straight trunks thrust skyward. From a distance the spire-shaped crowns make lodgepole forests look similar to spruce-fir forests, but the yellowish green color of the foliage is easily distinguished from the darker green foliage of spruces and firs, and lodgepole pines occur at lower elevations. Closer up, the needles are in bundles of two, and they are much shorter than those of ponderosa pine. Lodgepole pinecones can be a bit lopsided and sit half erect on their branches. Its scaly bark is rather thin, making lodgepole pines more susceptible to fire than other trees.

Rocky Mountain lodgepole pine (*Pinus contorta* var. *latifolia*) covers vast areas of the landscape where it forms remarkably pure stands of trees all of the same age. One wonders how this happens and the answer is: it is event- and ecology-driven. In fire-impacted landscapes, an adaptation called *serotiny* is at work. Cones are literally

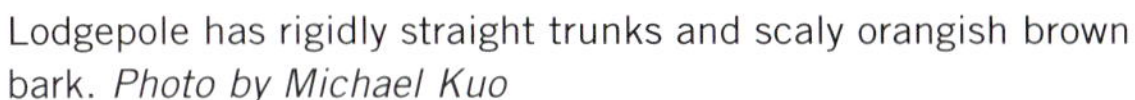
Lodgepole has rigidly straight trunks and scaly orangish brown bark. *Photo by Michael Kuo*

Lodgepole pines have short needles that are bundled in twos. *Photo by Michael Kuo*

glued shut with a sticky resin, sometimes for years until the heat of a wildfire melts the resin and releases the seeds all at once. The seeds then germinate, sometimes in tremendous numbers, over the landscape. Lodgepole pine is a pioneering tree and its seedlings do not tolerate shade. It is only when canopy trees are destroyed by wildfire or other disturbances and the forest floor is flooded with the light that the seedlings can flourish.

The seedlings grow synchronously upward toward the sun and in the process shade out tardy seedlings, producing thick "doghair lodgepole." Occasionally, a disturbance leaves a few older trees alive and a two-tiered forest is born, composed of sparse older trees reaching skyward with dense clusters of youngsters at their feet. The average lifespan of Rocky Mountain lodgepole pine is 150–200 years; in some places they are eventually replaced by shade-tolerant spruce and fir.

Under the lodgepole canopy, needleless but still living, flexible branches form a nearly impenetrable tangle, and the layer of duff is very deep. The forest floor is almost empty, but for a few scattered montane plants such as heart-leafed arnica (*Arnica cordifolia*), geraniums, snowbrush *Ceanothus*, and pasque flowers (*Anemone*

Lodgepole pine regenerating in a burned forest.
Photo by Cathy L. Cripps

Heart-leafed arnica under lodgepole pine.
Photo by Peter Lesica

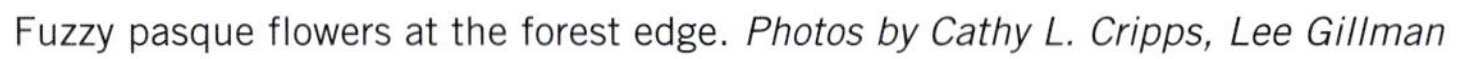

Fuzzy pasque flowers at the forest edge. *Photos by Cathy L. Cripps, Lee Gillman*

Fireweed in full bloom.
Photo by Loraine Yeatts

American red squirrels have reddish gray coats, white underbellies, and a white eye ring. *Photo by Michael Kuo*

A squirrel cache of lodgepole pinecones (notice the dried mushroom). *Photo by Cathy L. Cripps*

patens). Shrubs can include kinnikinnick (*Arctostaphylos uva-ursi*), gooseberry (*Ribes* sp.), and low juniper (*Juniperus communis*), which produces frosted blueberries in fall. The "frosting" is a wild yeast fungus that can be used as sourdough starter, if it is the correct strain. An abundance of fireweed often flourishes after fire (see next chapter, "Burned Ground," Pg 142).

Transient creatures such as deer, elk, and bear pass through but there is not much for them to eat in a lodgepole pine forest. Red squirrels (*Tamiasciurus hudsonicus*) eke out a living eating pine seeds; they also pile the cones in huge caches for winter. The squirrels seem to take delight in scolding intruders from treetops. Birds are not abundant in lodgepole pine forests so they can be quiet places. The white-breasted nuthatch (*Sitta carolinensis*) can be occasionally observed doing acrobatics as it hops about on bark looking for insect larvae and eggs. The colorful feathers of the pine grosbeaks (*Pinicola enucleator*) add a reddish accent to the forest scene. The similarly colored red crossbill (*Loxia curvirostra*) has an adaptation that helps it compete with the squirrels for supper. Its unique crossed beak is used like a can opener to pry open closed lodgepole pinecones to access the hidden seeds. Pine martins survive by hunting squirrels and birds in the forests, and porcupines subsist on the inner bark of lodgepole and other conifers in winter.

The red crossbill uses its beak to open lodgepole cones. *Photo by Stephen R. Jones*

Lodgepole pines are susceptible to mountain pine beetle (*Dendroctonus ponderosae*) infestations, which occur in natural cycles of boom and bust. In the last few years, large numbers of trees in the Rockies have been infected and miles and miles of red-brown, dead pines can be observed from an airplane. The tiny beetles, the size of a grain of rice, burrow into trees, creating "galleries," forming a characteristic pattern on the inside of the bark. The beetles carry hitchhikers called "blue stain fungi" (*Grossmania clavigera* and others) that proliferate in the tunnels. The fungi clog up the trees' plumbing, which eventually kills them. Larvae of the beetles eat the fungi and disperse their spores as adults in a symbiosis that benefits both organisms. The blue-stained wood is valued as lumber as well as for making interesting furniture.

Rocky Mountain lodgepole pines stretch from Canada to Colorado and much of its important range is in southwestern Montana, Idaho, Wyoming, and northern Colorado. Pure stands of 300- to 400-year-old trees exist in Yellowstone National Park. The park is one of the best places to view lodgepole pine regeneration because it can be found in various stages of recovery after sequential fires, including the massive 1988 Yellowstone Fires. In Colorado, the eastern slopes of the Front Range support persistent lodgepole pine forests.

While the mushroom flora of lodgepole pine forests is not as diverse as in many other Rocky Mountain forest types, many mushrooms are regularly found under lodgepoles—and a few, including *Tricholoma magnivelare*, *Suillus tomentosus*, and *Suillus brevipes*, are found almost exclusively in a mycorrhizal relationship with lodgepole. Some, such as the American matsutake (*Tricholoma magnivelare*) may be found only in forests of a certain age in late summer and fall. Springtime in the Rockies brings the "snowbank fungi" to lodgepole pine forests (Pg 188) and after fire, burn fungi appear (Pg 142). Midsummer droughts usually preclude fungal fruiting from late June through July, but with monsoonal moisture comes the autumn harvest, which can be prolific with *Russula* and *Lactarius* species and bounties of golden chanterelles (*Cantharellus roseocanus*).

Photo by Michael Kuo

Cystodermella cinnabarina (Albertini and Schweinitz) Harmaja

DESCRIPTION: CAP 3–8 cm across; egg-shaped or convex at first, becoming broadly convex, broadly bell-shaped, or nearly flat; dry; covered with mealy, granular scales; cinnabar-red to orange or rusty cinnamon. **GILLS** attached to the stalk but pulling away from it by maturity, close or crowded; white; at first covered by the partial veil. **STALK** 3–6 cm long; up to 1.5 cm thick; more or less club-shaped; dry; smooth and whitish to pale cinnamon near the apex, but sheathed with cinnabar granular scales from the base upward, the sheath terminating in a flimsy ring zone; granules often wearing away as the mushroom matures, exposing a coarse, whitish surface below. **FLESH** whitish; odor and taste slightly mealy, or not distinctive. **SPORE PRINT** white, not amyloid.

ECOLOGY: Saprobic; growing alone, scattered, gregariously, or in loose clusters; summer and fall. Often found in needle duff or in moss. Reported from Colorado, New Mexico, Wyoming, and Montana in the Rockies.

OBSERVATIONS: The granular orange cap and distinctively sheathed stem are hallmarks for this species. It has also been called *Cystoderma terryi*. It is difficult to distinguish this species from *Cystodermella granulosa* without a microscope; *C. cinnabarina* has spear-shaped cystidia on the gill edges, which the other species lack.

Photo by Cathy L. Cripps

Laccaria bicolor (Maire) P. D. Orton

DESCRIPTION: CAP 1–5 cm across, convex to plane, sometimes with a depression in the center; finely roughened; orange, orange-brown, pinkish brown; margin not striate, uplifted and wavy with age. **GILLS** attached, thick, well-separated, pink, sometimes with a lavender cast. **STALK** 2–7 × 0.5–1 cm, equal, or slightly larger at base; rough fibrous, cap color or lighter; basal mycelium evident, either whitish or lavender, fading. **FLESH** pinkish; firm and tough in stalk; odor not distinct. **SPORE PRINT** white, spores spiny.

ECOLOGY: Scattered in large groups over the forest floor in lodgepole pine and mixed conifer forests in summer and fall; mycorrhizal; known from the western United States and northern states.

OBSERVATIONS: *bicolor* for its lavender and pink colors. This common species of the Rocky Mountains is often misidentified as *Laccaria laccata* var. *pallidifolia*, which has a smoother cap and stalk. Confusion arises because the diagnostic lavender mycelium at the stalk base in *L. bicolor* can fade to white. *Laccaria nobilis* is similar with lavender tones, but it is not as common; it tends to occur at higher elevations. Both can be confused with pink-gilled Entolomas (pink spore print) or thick-gilled *Hygrophorus* species (decurrent gills).

Photo by Ed Barge

Lactarius rufus (Scopoli) Fries

DESCRIPTION: CAP 4.5–9 cm across, shallow convex, with or without a small bump in the center, smooth; dry; with a hoary coating at first, becoming deep red-brown to orange-red brown, often lighter toward margin; margin incurved at first. **GILLS** go slightly down the stalk; cream, becoming orange-buff where damaged; oozing white milk that does not change color when cut. **STALK** 4–5 × 1–1.5 cm, equal, straight; dry, smooth; with hoary coating at first, becoming more or less the color of the cap; odor mild; taste very hot. **FLESH** dingy white, discoloring brown. **SPORE PRINT** buff, spores with amyloid warts.

ECOLOGY: Occurring scattered or in clusters under lodgepole pine and mycorrhizal with this tree species; also found with other conifers; fruiting in summer and fall; widespread in the Rocky Mountains, northern states, and Canada.

OBSERVATIONS: *rufus* for the cap color; there are a number of similar-looking *Lactarius* species in Rocky Mountain conifer forests, but the others are more orange-brown and their taste is not as hot. The mushrooms in the photo are molecularly almost identical to the European *L. rufus*. Species that taste hot when raw should not be eaten.

Photo by Cathy L. Cripps

Russula brevipes var. *acrior* Shaffer

DESCRIPTION: CAP 5–10 cm across or larger, concave, dished in the center; with dry kidskin texture, soil adhering; dingy white, staining yellow-brown; margin turned under at first. **GILLS** narrowly attached, narrow, crowded, some dividing at the stipe; whitish with a pale mint-green cast, staining brown. **STALK** buried in the duff; 2–5 × 2–3 cm, short and squat, equal or pointed at the base; smooth or slightly pubescent; dingy white, often with a thin greenish line at the apex, staining brown. **FLESH** white, yellow-brown in maggot holes, very firm, almost hard; odor indistinct or unpleasant; taste hot. **SPORE PRINT** creamy.

ECOLOGY: Mycorrhizal with lodgepole pine and possibly other conifers; gregarious, very common late in the fall from August to October, often one of the last mushrooms to fruit in the Rocky Mountains; the very large caps are typically almost buried in the duff as "mush lumps."

OBSERVATIONS: *brevipes* for its short stipe and *acrior* for its acrid taste. The gills run down the stalk making it look like a *Lactarius* but it lacks milky latex. *Russula brevipes* is one of a group of dense-fleshed Russulas; others such as *R. nigricans* and *R. albonigra* turn blackish. The variety *acrior* has a greenish cast and is very hot, so it is not an edible fungus.

Photo by Michael Kuo

Tricholoma focale (Fries) Ricken

DESCRIPTION: CAP 2.5–10 cm across; convex, becoming broadly convex to nearly flat when mature; sticky at first, but soon dry; covered with long fibrils; brown to yellow-brown or, more commonly, orange-brown; often developing olive hues; the margin at first inrolled. **GILLS** attached to the stalk by means of a notch; close or crowded; whitish; often discoloring and spotting brown. **STALK** 4.5–10 cm long; 1–3 cm thick; more or less equal, or tapering toward the base; with a cottony white ring that discolors brownish and often collapses with age; smooth above the ring, but shaggy below it, with brown fibrils over a buff ground color. **FLESH** whitish or slightly brownish in places; not changing on exposure; odor and taste mealy. **SPORE PRINT** white.

ECOLOGY: Mycorrhizal, in conifer woods, particularly lodgepole pine in the Rocky Mountains, also found in other areas of the West; growing alone, scattered, or gregariously; fall.

OBSERVATIONS: The orangish brown cap that develops yellow and olive shades, the collapsing cottony ring, and the mealy odor are good field characters for this species. Previous names include *Tricholoma zelleri* and *Armillaria zelleri*. Caution is advised as some Tricholomas are poisonous.

Photo by Vera S. Evenson

Tricholoma magnivelare (Peck) Redhead

DESCRIPTION: CAP 5–10 cm across; convex, becoming broadly convex or nearly flat; dry or a little sticky; white or beige at first, but soon with brownish discolorations, scales, and pressed-down fibers; the margin cottony and rolled under when young. **GILLS** attached to the stalk, sometimes by means of a notch; close or crowded; ivory white, developing brown or reddish brown stains and spots with age. **STALK** 4–10 cm long; up to 2.5 cm thick; more or less equal, or with a slightly tapered base (but not with a long, rooting base); white above the ring, but colored like the cap below; partial veil white and thick, collapsing to form a sheath around the lower stem and a prominent flaring ring at the top edge of the sheath. **FLESH** white; firm; not changing on exposure; taste spicy; odor fragrant and distinctive (cinnamonlike and spicy, but earthy as well). **SPORE PRINT** white.

ECOLOGY: Mycorrhizal with pines; growing scattered or gregariously; late summer and fall. Reported from Colorado, New Mexico, Wyoming, and Idaho in the Rocky Mountain region.

OBSERVATIONS: This highly prized edible is hard to spot when young since it appears as a mush lump, but it can sometimes be sniffed out and easily recognized by its strong and characteristic spicy odor. In the Rocky Mountains, it appears to be exclusively associated with lodgepole pines—though it is well documented with other trees in other geographic areas. Look-alikes include *Tricholoma focale* (Pg 131), which has a mealy odor, and *Catathelasma* species, which have a double ring and a mild to mealy odor. Known as the American matsutake.

Photo by Vera S. Evenson

Stropharia hornemannii (Fries) S. Lundell and Nannfeldt

DESCRIPTION: CAP 4–16 cm across; convex, becoming broadly convex to flat or broadly bell-shaped; sticky when fresh; bald; reddish brown to purple-brown, brown, or olive-brown when fresh, but often fading to tan or pale yellowish brown; occasionally decorated with white veil remnants near the margin. **GILLS** attached to the stalk or beginning to pull away from it; pale gray at first, becoming purplish gray to purple-black; close. **STALK** 5–15 cm long; up to 2.5 cm thick; equal; dry; conspicuously shaggy, especially when young; whitish; decorated with a skirtlike, white ring that becomes dusted with purple-black spore powder; base attached to white mycelial threads. **FLESH** white; odor and taste not distinctive, or slightly unpleasant. **SPORE PRINT** dark purple-brown to black.

ECOLOGY: Saprobic on duff and woody debris of various conifers; growing alone, scattered, gregariously, or in clusters; late summer and fall, widely distributed.

OBSERVATIONS: This charismatic species tends to grow in clusters and is often found fruiting from well-decayed wood. Some *Pholiota* species also have shaggy stalks and tissue on their margin and grow on wood, but they have brown spores.

Photo by Michael Kuo

Suillus brevipes (Peck) Kuntze

DESCRIPTION: CAP 5–10 cm across; convex, becoming broadly convex and remaining so for a long time—or eventually becoming more or less flat; slimy; bald; dark reddish brown, fading to cinnamon or yellowish brown; the margin at first incurved and pale but naked (without veil remnants). **PORE SURFACE** pale yellow, becoming dingy olive, not staining; 1–2 circular pores per mm; tubes to about 1 cm deep. **STALK** 2–5 cm long; 1–3 cm thick; swollen and squat when young; often short, even at maturity; white at first, becoming pale yellow; glandular dots absent; without a ring. **FLESH** white at first, becoming yellow with age; soft; not staining when sliced; odor and taste not distinctive. **SPORE PRINT** brown to dull cinnamon.

ECOLOGY: Mycorrhizal with at least lodgepole pine in the Rockies; growing alone, scattered, or gregariously; late summer and fall. Often found in young lodgepole stands, a few years after a forest fire has gone through.

OBSERVATIONS: *Suillus brevipes* is similar to *Suillus granulatus* and *Suillus kaibabensis* (Pg 83) but can usually be distinguished by its squat smooth stalk and long-remaining-convex cap, along with its association (in the Rocky Mountains) with lodgepole pine. *S. brevipes* also lacks the glandular dots that are obvious on the stalks of the other *Suillus* species. Edible, but the cap should be peeled.

Photo by Vera S. Evenson

Suillus tomentosus (Kauffman) Singer

DESCRIPTION: CAP 5–15 cm across; convex, becoming broadly convex; sticky or fairly dry; at first covered with a fine, grayish, felty covering, but often becoming smoother with age; yellow to orangish yellow; sometimes developing reddish spots and stains; the margin at first inrolled. **PORE SURFACE** brownish to cinnamon when young, becoming brownish yellow to olive-yellow; bruising blue; 1–2 angular pores per mm; tubes to 2 cm deep. **STALK** 4–12 cm long; 1–3 cm thick; equal or somewhat club-shaped; yellowish orange; covered with brownish glandular dots; staining brownish on handling; without a ring. **FLESH** whitish to yellowish in the cap; yellow in the stem; bluing on exposure; odor and taste not distinctive. **SPORE PRINT** olive-brown to cinnamon-brown.

ECOLOGY: Mycorrhizal with lodgepole pine; growing scattered or gregariously; summer and fall; one of the most common *Suillus* species in this habitat, it can be exceedingly abundant at times.

OBSERVATIONS: The distinctive felty orange cap and the bluing flesh are key identifying features for this ubiquitous lodgepole companion. The variety *discolor* (Pg 217) has a darker cap and is associated with 5-needle pines (see High Pine Forests). Some people eat this species, but it is not a choice edible.

Photo by Vera S. Evenson

Rhizopogon ochraceorubens A. H. Smith

DESCRIPTION: SPORE CASE 2–6 cm broad, rounded to somewhat oblong; exterior golden yellow to tawny-ochraceous covered by distinctive reddish brown, net-like rhizomorphs, staining brownish red where injured. **STALK** absent. **FLESH** minutely chambered (visible with lens); off-white, firm and crisp in young, becoming pale olive with age; odor mild, taste not reported.

ECOLOGY: Mycorrhizal, under lodgepole pines, less commonly with ponderosa pines; fruiting underground and often erupting through the conifer duff in late summer; often several fruiting bodies are found just above the mineral soil level. Widely distributed in the Rocky Mountain region.

OBSERVATIONS: Many *Rhizopogon* species are found with lodgepole and other pines and they are difficult to tell apart. These hypogeous mushrooms develop underground, maturing their spores inside a spore case. The problem of spore dispersal is usually solved by the production of strong odors in the mature fungi so that small foraging animals such as squirrels find them and dig them up for a nutritious treat. The spores are adapted to pass through the animal's gut unharmed and be dispersed in the feces, perhaps far from the original fruiting.

Photo by Ed Barge

Cantharellus roseocanus (Redhead, Norvell, and Danell) Redhead, Norvell, and Moncalvo

DESCRIPTION: FRUITING BODIES: CAP 3–12 cm across; more or less plano-convex when young (often with an inrolled margin), becoming flat or shallowly depressed, with a wavy and irregular margin; tacky when wet but soon dry; pale yellow to egg-yolk yellow or orange when fresh, but often fading to very pale yellow or nearly whitish when exposed to sunlight; with a pale to dark pink bloom when young, especially near the margin. **FALSE GILLS** well developed; running down the stem; frequently cross-veined; bright, intense orange (usually contrasting markedly with the cap surface). **STALK** 2–5 cm long; up to 2.5 cm thick; variable in shape but often stocky; smooth; colored like the cap before it fades or colored like the false gills. **FLESH** whitish; unchanging when sliced; solid; odor fragrant and sweet, reminiscent of apricots; taste mild to slightly fruity. **SPORE PRINT** pale orange-yellow.

ECOLOGY: Mycorrhizal; growing alone, scattered, or gregariously, in lodgepole pine or spruce-fir forests; summer and fall; distributed throughout the Rockies.

OBSERVATIONS: The Rocky Mountain or rainbow chanterelle is one of the best edible mushrooms in North America—superior in texture and flavor to all other North American chanterelles we have tried. Hallmark features include the bright orange false gills (that are not structurally distinct and thus cannot be separated easily from the mushroom), which contrast with the duller cap surface, and the fruity odor, best detected when several to many mushrooms have been sitting in a basket or bag together for a few hours.

Photo by Ed Barge

Fomitopsis pinicola (Swartz) P. Karsten

DESCRIPTION: CAP up to about 40 cm across and 10 cm deep; semicircular or fan-shaped in outline; convex or hoof-shaped; bald and lacquered in appearance when young (or in fresh annual zones), maturing to become wrinkled and cracked; red to dark brownish red, with a white to yellow or reddish marginal zone—but often becoming grayish brown overall after several years. **PORE SURFACE** cream; not bruising; with 3–6 round pores per mm; tube layers usually fairly distinct, layers annually produced, up to 8 mm deep. **STALK** absent. **FLESH** whitish; leathery to woody; odor strong and musty when fresh.

ECOLOGY: Saprobic on the deadwood of conifers and, sometimes, hardwoods; also sometimes parasitic on living trees; causing a brown cubical rot; growing alone or gregariously; perennial.

OBSERVATIONS: This decomposer of dead conifer wood is essential to many North American ecosystems, including lodgepole pine forests. It is easily recognized by its vibrant red or orange "belt" when young.

Photo by Vera S. Evenson

Gloeophyllum sepiarium (Wulfen) P. Karsten

DESCRIPTION: CAP single or compound (and then either fused laterally or with loosely arranged lobes arising from a central point); up to about 12 cm across and 8 cm deep; semicircular, irregularly bracket-shaped, or kidney-shaped in outline; flattened-convex; velvety to hairy; rugged; with concentric zones of texture and color; at first yellow to orange, becoming yellow-brown to dark brown or nearly black toward the point of attachment—but usually remaining yellow to orange on the growing margin. **FALSE GILLS** irregular, slotlike, and often fusing; fairly close; often mixed with elongated pores; edges yellow-brown, becoming darker brown with age; faces cream to pale brownish, darkening with age; up to about 1 cm deep. **STALK** absent. **FLESH** dark rusty orange to brown or dark yellow-brown; corky; odor not distinctive.

ECOLOGY: Saprobic on the deadwood of conifers and, sometimes, hardwoods; causing a brown rot; growing alone or gregariously; annual, or sometimes reviving to be perennial. Widely distributed in Rocky Mountain regions.

OBSERVATIONS: The orange to yellow growing margin, the slotlike false gills, and the rusty orange to brown fibrous flesh are good field characters for this commonly encountered species.

Photo by Cathy L. Cripps

Gyromitra esculenta (Persoon) Fries

DESCRIPTION: CAP 4–8 cm high × 4–10 cm wide, irregularly shaped, reminiscent of a walnut, composed of numerous wrinkles and folds; surface smooth, greasy; reddish brown; margin overlaps the stalk. Underside smooth, lighter in color. **STALK** 3–6 × 1.3–3 cm, equal or wider at top or bottom, often flattened with a longitudinal indent but not ribbed; puckered at the base; smooth or slightly bumpy; cream with a pinkish brown tinge. **FLESH** rubbery but brittle; stuffed or hollow in stalk; not chambered; odor not distinct. **SPORES** light colored.

ECOLOGY: Scattered in lodgepole pine and also in mixed forests in the lower subalpine, but not yet confirmed as mycorrhizal. Reported from all latitudes of the Rocky Mountains and widely distributed throughout North America; fruiting in spring and early summer.

OBSERVATIONS: *esculenta* is a misnomer for this poisonous species. It contains gyromitrin, which becomes monomethylhydrazine (a type of rocket fuel) in the body. Eating too many over time can cause gastrointestinal distress, loss of coordination, headache, and possible kidney and/or liver failure since the body accumulates the toxin. The toxin is volatile and it is often the cook who gets sick; deaths have occurred in Europe. *Gyromitra esculenta*, known as the "false morel" has been mistaken for *Gyromitra montana* (Pg 205) and should definitely be avoided.

Photo by Michael Kuo

Helvella acetabulum (Linnaeus) Quélet

DESCRIPTION: CAP 1–8 cm; cuplike; sometimes becoming flat with age; upper surface brown to yellowish brown, bald; undersurface brown to yellow-brown, sometimes paler near the stalk, finely hairy near the margin, with forked ribs extending from the stalk, sometimes almost to the margin; the margin more or less even. **STALK** to 9 cm long and 3 cm thick; becoming broader near the cap; deeply ribbed with sharp-edged ribs that extend far onto the undersurface of the cap; cream. **FLESH** thin; brittle. **SPORE PRINT** white.

ECOLOGY: Probably mycorrhizal; growing alone or gregariously under hardwoods or conifers; spring and early summer; widely distributed.

OBSERVATIONS: This fascinating cup fungus has an amazing, ribbed stem; the ribs extend elaborately onto the undersurface of the cap. Other brown cup fungi lack this feature. Not edible.

Photo by Cathy L. Cripps

BURNED GROUND

Wildfire, as an integral part of the western landscape, consumes thousands of acres of forest and grassland every year in the Rocky Mountains. And yet, from what appears to be a force of death and destruction, comes renewal and regeneration. In summer, skies darken and roily clouds accumulate as mountain storms roll in to release their moisture. At higher elevations, hikers feel their hair stand on end in the electrical buildup before lightning releases its energy and thunder closes the vacuum with massive booms and a downpour begins. Lightning strikes trees and smolders dried duff, which can burst into the bright orange flames of ground and crown fires; driven by wind these fires can run for miles producing massive plumes of smoke. Wildlife flees or is burned but the flames are inescapable to plant life. The aftermath is a blackened landscape of ash, charcoal, rocks, and tree cadavers.

Then slowly, quietly, life returns. Seedlings of all kinds spring from the blackened carpet. Fire-adapted plants, with seeds that need heat to germinate may have waited years for this opportunity. Plants such as redstem Ceanothus (C. *sanguineus*) have

The Derby fire produced this extensive burn in Montana.
Photo by Cathy L. Cripps

Magenta-colored fireweed regenerating in a burned-over habitat.
Photo by Andy Hogg

heat-resistant seeds that germinate after fire. Fireweed (*Chamerion angustifolium*) and pine grass (*Calamagrostis rubescens*) are fire-stimulated species that resprout from rhizomes and flourish in a fire's aftermath.

Certain western trees are also adapted to the scenario of periodic wildfire. Lodgepole pines need to burn in order to regenerate themselves. Heat melts the sticky resin covering their "serotinous" cones and releases seeds by the thousands all at once. The results after a few years can be acres of "doghair" lodgepole, which stand shoulder-to-shoulder as the young trees develop into new forests. Aspen also

Whitebark pine seedling germinating.
Photo by Cathy L. Cripps

Lodgepole pine regenerating as thick "doghair" stands after fire.
Photo by Cathy L. Cripps

regenerates after fire and their ramets sprout by the hundreds from root systems that are not totally burned. Also, aspen mother trees produce seeds when stimulated by fire. Whitebark pine benefits from wildfire because shade-tolerant spruce and fir can encroach on its territory over time. Fire clears the landscape so pioneering sun-loving whitebark pines can return via the nutcrackers, birds that plant their seeds in burns.

The animals soon return, and browsers like deer, elk, and moose go after young sprouts for a tasty treat that is available in abundance. Cavity-nesting birds such as woodpeckers find the burned snags irresistible. The American three-toed woodpecker (*Picoides dorsalis*) and the black-backed woodpecker (*Picoides arcticus*) are specialists that prefer burned forests, and their black-and-white coloration matches the landscape. Likely they come because beetle populations can surge in burned forests. Bark beetles and wood-boring beetles lay their eggs in seared trees that cannot pitch them out. Their fat juicy larvae nestled in the snags are sought out by woodpeckers and other birds. As the diversity of plant life returns, so do butterflies and other pollinators. They are sometimes observed in record numbers a year to two after fire has swept through.

Just as there is plant succession after wildfire, so too is there a fascinating story of fungal succession. Fire releases nutrients into the soil and the burned ground is basically a blank slate devoid of organisms. Those fungi able to move in fast may not be the best competitors later on, but they thrive in the fire-forged black soil. Interestingly, it is mostly ascomycetes, the morels and cup fungi that have taken advantage of this space. Depending on the severity of the fire, one and sometimes two or three years after a fire, an abundance of tiny cup fungi such as *Geopyxis*

Elk grazing in a burn and a three-toed woodpecker nesting in a burned tree. *Yellowstone National Park Photos*

Burn Pholiotas fruiting in an old campfire pit. Mycelium running through burned soil. *Photos by Don Bachman*

carbonaria, *G. vulcanalis*, and *Anthracobia* species can be observed covering the ground; their names attest to their burn-loving nature. Their delicate mycelium may help stabilize the fragile soil after fire.

Morels appear, often by the thousands, sometimes densely packed and also dispersed over acres and acres. Burn morels are often most prolific one year after a fire, but fruiting can persist for several seasons. These abundant fruitings might be a result of the uncontested space devoid of competitors, fast-growing mycelium, spore germination stimulated by heat, or special soil nutrients released by fire, but the phenomenon is not well understood. There are six known morel species adapted to burns, some of which can be distinguished only by their DNA.

The first flushes of cup fungi and morels are followed by the appearance of saprobic basidiomycetes, many of which fruit only in burns, whether it is an extensive burn or only an old campfire pit. A number of "burn Pholiotas" have adapted to this habitat, as have a few species of *Psathyrella*, *Tephrocybe*, and *Myxomphalia*. These saprobic species are eventually followed (a few years after a fire) by fungi that are mycorrhizal with the new tree seedlings, the so-called "early colonizer fungi." After the great Yellowstone fires of 1988, *Inocybe lacera* and *Laccaria* species were prolific around young lodgepole pine seedlings, as were *Suillus brevipes* and *Coltricia perennis*.

Burn morels in conifers needles and a single giant burn morel (*M. septimelata*). *Photos by Ed Barge, Don Bachman*

Commercial pickers have taken advantage of morel burn ecology and have become as prolific as the fungi in recent years. This has resulted in more regulation, and permits are now necessary for commercial as well as casual pickers collecting burn morels on public lands in most Rocky Mountain states. The best sources of information are often USDA Forest Service Web sites or offices. However, for the casual picker, it is best to stay away from areas designated as "commercial" because of the competition. Look for burns, one or two years after a fire, primarily in areas previously forested in lodgepole pine and spruce-fir. Land previously in ponderosa pine or Douglas fir is also worth a look. If there are a lot of pickers in an area, the farther you walk or the more willing you are to access difficult areas, the more likely you are to find overlooked morels. Even small burns can produce morels as well as a sufficient supply of intriguing burn fungi to keep a collector busy. Don't overlook old fire pits for an occasional fungal surprise.

Photo by Cathy L. Cripps

Tephrocybe atrata (Fries) Donk

DESCRIPTION: CAP 1–2.5 cm, convex, with a small nipple or sunken center; dark umber-brown, fading to brown; smooth, greasy; margin turned down and with faint lines. **GILLS** attached, well-separated; whitish cream at first, becoming pale brown. **STALK** 1–4 × 0.3–0.4 cm, thin, equal, smooth; cream to pale brown, but can be darker. **FLESH** white; odor mealy. **SPORE PRINT** white, spores elliptical.

ECOLOGY: Scattered on the ground in large burns, slash burns, and old campfires; a decomposer of burned organic material; this species is likely common in the Rocky Mountains but is underreported.

OBSERVATIONS: One of the few gilled mushrooms with white spores found in burns. Another is *Myxomphalia maura*, which is more funnel-shaped, gray, with crowded gills that run down the stalk. Rocky Mountain burn Tephrocybes are often reported as *T. anthracophila*, which has round spores. The mushrooms in the photo had elliptical spores like *T. atrata*.

Photo by Cathy L. Cripps

Pachylepyrium carbonicola (A. H. Smith) Singer

DESCRIPTION: CAP 1.5–2.5 cm across, convex, smooth; at first covered with whitish fibrils, then date-brown with a white band on margin. **GILLS** attached; pale orange, then orange-brown. **VEIL** cobwebby; observed mostly as fibrils on the stalk; often covered with rusty spores. **STALK** 3–5 × 0.3–0.4 cm, thin, equal; white, with white fibers below ring zone. **FLESH** cream or pale brown; odor indistinct. **SPORE PRINT** rusty red-brown.

ECOLOGY: A decomposer found on burned ground, fruiting in groups, often near melting snowbanks in spring. Known from the Northern Rockies and the Pacific Northwest, and likely in the Southern Rockies; apparently rare or rarely recognized.

OBSERVATIONS: *carbonicola* for its habitat. This species has also been called *Pholiota subangularis* (Smith and Hesler 1968) because some spores are slightly angular, but it is the bright rusty spore color that is striking and diagnostic. This little burn fungus looks like a cross between a *Cortinarius* (cobwebby veil and rusty spores) and a *Pholiota* for its stature; another name is *Crassisporium funariophilum*. When young, it is completely covered with white hairs giving it a frosted appearance. Not edible but very beautiful.

Photo by Cathy L. Cripps

Pholiota highlandensis (Peck) Quadraccia and Lunghini

DESCRIPTION: CAP 2–3.5 cm across, convex, sometimes with depression in center, smooth, lubricous, sticky; bright orange-brown with a slight yellow frosting toward the margin; margin turned down or in, covered with a sticky veil at first. **GILLS** attached, narrow; cream, pale gray, buff, later orange-brown. **VEIL** pale brown. **STALK** 2–5 × 0.5–1 cm, narrowing toward base, some flattened; white or cream, later with yellow tints, with scattered brown tufts below the ring zone. **FLESH** white, yellow under the cap skin, black in the stipe base; odor not distinctive. **SPORE PRINT** cinnamon-brown.

ECOLOGY: A decomposer on burned soil, often near burned trees, stumps, or logs, sometimes fruiting in huge numbers. It is best observed in the Rockies in spring near melting snowbanks at high elevations in the subalpine zone, but it also occurs at lower elevations on burns and in fire pits; one of the most common "burn fungi" in the Rockies. Widely distributed.

OBSERVATIONS: *highlandensis* for its original location in Highland Falls, N.Y. There are several Pholiotas that prefer burns, including *P. fulvozonata*, which has a reddish veil and red fibrils down the stipe, and *P. molesta* and *P. brunnescens* (Smith and Hesler 1968), which have brown caps. *Pachylepyrium carbonicola* (Pg 148) is red-brown with a thinner stipe. Edibility unknown.

Photo by Cathy L. Cripps

Psathyrella pennata (Fries) A. Pearson and Dennis

DESCRIPTION: CAP 1–3(4) cm across, convex, bell-shaped; red-brown at first, becoming pale brown, grayish brown, or yellowish brown; covered with white hairs especially at the margin giving it a shaggy appearance; hairs less obvious on aging or after rain. **GILLS** attached, crowded; gray-brown or brown, edges white. **VEIL** pale brown, remaining as tissue on cap margin. **STALK** 2–5 × 0.2–0.4 cm, long and thin, equal; whitish and covered with white fibrils; ring absent. **FLESH** thin, buff; odor not distinctive. **SPORE PRINT** blackish.

ECOLOGY: A saprobe occurring in clusters on or near burned wood in large or small burns; often fruiting in spring but found throughout the summer; it has a wide distribution in North America and Europe, wherever burns occur.

OBSERVATIONS: called *Psathyrella carbonicola* by Alex Smith for its habitat. The clustered habit, blackish spores, association with burned (sometimes buried) wood, and white hairy tissue on the cap help confirm this species. When young, caps are completely covered with white hairs that are eventually reduced to bits of tissue mostly on the cap margin. *Pachylepyrium carbonicola* also has white tissue on the cap but has rusty red-brown spores.

Photo by Cathy L. Cripps

Geopyxis carbonaria (Albertini and Schweinitz) Saccardo

DESCRIPTION: FRUITING BODY 1–2 cm across, a deep urn-shaped cup; interior smooth; dull to bright orange-brown; exterior a bit rough, also orange-brown; with a rim of white ragged tissue. **STALK** 1–3 × 0.2–0.5 cm, thin, short or long, whitish. **FLESH** brittle, easily breaking; odor not distinctive. **SPORES** whitish.

ECOLOGY: Sometimes fruiting in large numbers over burned ground. The most common cup fungus on burns in the Rockies, this species occurs in spring, often just before or during morel season in May and June, but it also appears later in the year. It is suspected of having an association with conifer roots that is possibly mycorrhizal.

OBSERVATIONS: *carbonaria* for its habitat. The deep cup shape, orange color, white edge, and stipe make this an easily recognizable burn cup. *Geopyxis vulcanalis* has a shallow, yellow-brown cup with white exterior and a sulfurlike odor when crushed between the fingers. Many of the burn cups are described in Beug et al., 2014. Anecdotal information suggests this fungus is a good indicator of where morels will fruit, but this has not been confirmed.

Photo by Cathy L. Cripps

Morchella septimelata M. Kuo

DESCRIPTION: CAP 4–8 cm tall, conic with a narrowing apex, or egg-shaped and more rounded; composed of pits and ridges in vertical rows, with cross veins; wood-brown, dark brown, black-brown, some with pinkish tints; pits light yellow-brown or pale brown; ridges dark brown velvety in young specimens but smoother and lighter with age; margin indented where it connects with stipe making a small white "track." **STALK** 1.5–3 × 1.5–2.5 cm, often constricted in middle, finely bumpy, with a few puckers; white. **FLESH** whitish, brittle; hollow; odor not distinctive.

ECOLOGY: Scattered or clustered on burned soil, sometimes fruiting in huge numbers one and two years after a burn; often found among the needles that have fallen off burned trees, such as lodgepole pine and other conifers, and possibly associated with conifer roots. Likely widely distributed throughout the Rocky Mountain region.

OBSERVATIONS: There are a number of "black burn morels" and their names are not yet fully determined. Molecular analysis matched this one to *M. septimelata* (= *M. eximia Boudier*), which cannot be distinguished from *M. sextelata* in the field. Burn morels are a high-quality edible, picked and sold commercially. However, caution is advised when considering any morel for the table. They should not be eaten raw and need to be cooked well; old specimens should not be consumed. Many people have an adverse reaction to cooked morels; a few cannot consume alcohol with morels without an adverse reaction. It is best to know your own personal tolerance for each kind of morel.

Photos by Taylor Lockwood, Cathy L. Cripps

Morchella tomentosa M. Kuo

DESCRIPTION: CAP 3–11 cm tall and 2–5 cm wide; egg-shaped or nearly conical; pitted and ridged, with the pits and ridges typically densely packed when young, stretching with maturity and developing vertical orientation; with gray to nearly black ridges and pits that are densely fuzzy when young; often developing grayish, pale tan, yellowish, or even whitish pits and ridges when mature; margin completely attached to the stalk; hollow. **STALK** 2–6 cm high and 1–4 cm wide; often swollen at the base; dark gray to nearly black and densely fuzzy when young; becoming pale (whitish to yellowish or grayish) with age, usually with stretched-out brownish patches of remaining fuzz; hollow.

ECOLOGY: Possibly saprobic and mycorrhizal at different points in its life cycle; growing alone, scattered, or gregariously in burned conifer forests, primarily in the year following the fire but sometimes appearing in dwindling numbers for a few years thereafter; spring (accounting for elevation). This morel appears late in the season, after other burn morels have passed their prime. Occurring in Western North America as far north as Alaska.

OBSERVATIONS: This distinctive morel is densely fuzzy and dark, at least when young and fresh. Several other burnsite black morels are common in the Rockies and feature smooth surfaces; they are very similar in their physical features and often cannot be definitively identified without DNA analysis. Caution is advised when eating burn morels, especially for the first time (Pg 152).

Photo by Shawna Crocker

Anthracobia melaloma (Albertini and Schweinitz) Boudier

DESCRIPTION: FRUITING BODY 3–15 mm in diameter, at first nearly globose, then opening to cup shapes; inner surface of cups strong yellow-orange, smooth; margins with brownish orange indentations; outer surface of cups somewhat lighter orange, sparsely roughened, clothed with bunches of brown hairs. **BASE** sessile or with very short stalk, attached deeply to charcoal and burned soil. **SPORES** whitish.

ECOLOGY: A burnsite fungus found in charcoal and cinders, sometimes by the thousands in conifer forests after wild fires have swept through; easily noticed because of the contrast in colors with the blackened soil, sometimes tiny green *Funaria* mosses can be seen growing nearby. Reported from the Pacific Northwest, Colorado, and Arizona. The fungi pictured here were found by school children in the Black Forest, El Paso county, Colorado, after the extensive ponderosa pine forest burn there in 2013.

OBSERVATIONS: Since wild fires are more and more prevalent in the Rocky Mountain region of late, we should expect to see these little pioneers peeking through the blackened debris more often. Old campfires are also an interesting place to inspect for burnsite fungi. There are several orange to red burnsite cups known from the Rockies, differing by external hairs, spore characteristics, and subtleties of fruiting body color. Members of the genus *Octospora* have similar disc-shaped orangish to reddish cups, but their outsides are smooth without hairs and their spores differ. *Pyronema omphalodes* is a fairly common burnsite fungus; it is distinguished by its often confluent clustered masses of pinkish orange to reddish orange cups growing on a dense whitish mycelial mat.

Photo by Steve Brace

Peziza sublilacina Svrček

DESCRIPTION: FRUITING BODY cup- to saucer-shaped, 1–3 cm wide; inside spore-bearing surface is smooth, lilac, maroon to reddish violet; outer surface pruinose roughened, pale grayish. **STALK** absent. **FLESH** pale violet, fragile; odor mild, taste not recorded. **SPORES** light-colored, smooth.

ECOLOGY: Gregarious to scattered on burned ground; spring and summer; not common but since wildfires have been occurring more often in the Rockies of late, these distinctive fungi should be expected to be found a year or two after a fire has moved through an area.

OBSERVATIONS: Previously known as *Peziza violacea*, but this name is confused in the literature. There are several burn fungi that have similar colors and shapes but must be differentiated on spore features and other microscopic characteristics. *Peziza praetervisa* is another violet-colored cup that is common on burns, but it is often larger and has finely warted spores (Beug et al. 2014).

Photo by Cathy L. Cripps

Plicaria endocarpoides (Berkeley) Rifai

DESCRIPTION: Fruiting body 5–13 cm across, a large irregular flattened cup; interior with large wrinkles or folds radiating from the center; dark reddish or violaceous brown with a distinct dark olive-green or blackish green coloration; exterior more brown or black-brown, with a fine bumpy texture. **STALK** absent. **FLESH** brittle, fragile, cartilaginous, but also can be rubbery. **SPORES** whitish or pale yellow.

ECOLOGY: Single or in crowded clusters on burned ground in subalpine and montane forests, wherever wildfires have occurred; fruiting in spring from May to June, but also appearing in summer; known throughout the Rockies and Europe, often associated with mosses on burns, but their ecological relationship is unclear.

OBSERVATIONS: This is one of the larger cup fungi found on burns and is recognized by its irregular shape and the purplish brown color tinged with olive-green inside the cup. Other *Plicaria* species on burns include *P. carbonaria*, which is dark brown to blackish and has spores with blunt, wartlike spines. *P. endocarpoides* spores are smooth in contrast to other species in the genus. Not edible and not to be confused with cuplike *Gyromitra* (*Discina*) species, which also have a wrinkled interior, but are more brown.

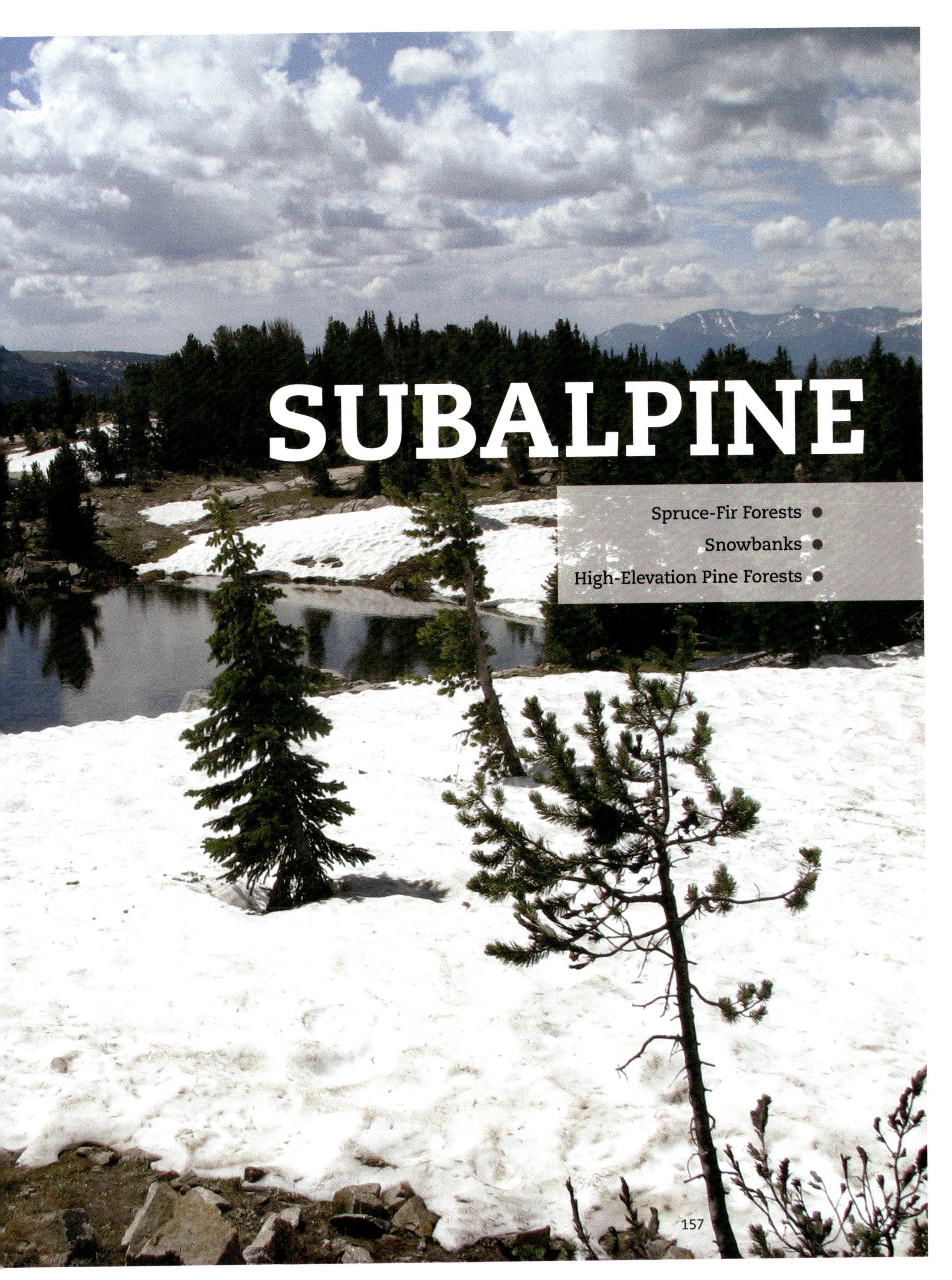

SUBALPINE

- Spruce-Fir Forests
- Snowbanks
- High-Elevation Pine Forests

Photo by Michael Kuo

SPRUCE-FIR FORESTS

In the upper elevations of the Rocky Mountains, spruce-fir forests spread downward from the tree line like a blanket over the peaks and valleys. Composed primarily of Engelmann spruce and subalpine fir (sometimes mixed with Colorado blue spruce in the Southern Rockies); the forests themselves are densely packed tangles of narrow, spire-shaped trees with interlocking branches, pitched for miles and miles on steep, rocky slopes; in short, spruce-fir forests are fairly impenetrable and are usually enjoyed by hikers and nature enthusiasts at their edges, where sunlit, grassy areas meet the woods. As North American tree icon Donald Culross Peattie puts it, the Engelmann spruce "comes crowding down to the edge of the meadow where your tent is pitched, to the rocks surrounding the little lake that mirrors its lancelike forms upside down."

Engelmann spruce (*Picea engelmannii*) and subalpine fir (*Abies lasiocarpa*) occur together so often that the habitat they form is simply known as "spruce-fir." Together

Stature of subalpine fir. *Photo by Michael Kuo*

they dominate the higher elevations on mountain slopes and form a krummholz of low, deformed trees at the tree line where high pines are absent. The scaly reddish bark of spruce contrasts with the smooth white bark of the fir. In reference to the needles, children are taught that "spruce is sharp" and "*fir* is *flat*" or "*friendly*." Fir cones often go unnoticed since they sit upright on the upper branches and often fall apart on the tree. Spruce cones, on the other hand, hang down, are papery, and remain intact on the ground. Vast areas of spruce-fir have been destroyed by bark beetles in the Southern Rockies and spruce bud worm has caused serious losses in the north.

The interior of a spruce-fir forest is a tangle of tipped trees piled tenuously like pickup sticks in various states of decomposition; but trails allow access where hiking is difficult. The forest floor is often covered with a low layer of *Vaccinium* and moss species. Here and there a ray of light might penetrate and illuminate such precious plants as fairy slipper orchids (*Calypso bulbosa*) and miniature red columbines (*Aquilegia elegantula*). It is fascinating to

Subalpine fir has smooth, whitish bark with gray blisters and flat needles that do not leave pegs when they fall. *Photos by Michael Kuo*

Engelmann spruce has distinctive scaly, reddish bark and papery cones. *Photos by Michael Kuo, Cathy L. Cripps*

think that the tiny seeds of lady slippers require certain mycorrhizal fungi from the forest to germinate. Jacob's ladder (*Polemonium pulcherrimum* spp. *delicatum*) brightens open areas with its deep purple-blue colors and surprises when its skunklike odor is inhaled. Low mats of strawberries (*Fragaria* spp.) can punctuate the forest floor and open areas of the subalpine can also be a place for picking mountain huckleberries and red raspberries.

Beetle kill, the result of bark beetles. Vast stretches of the spruce-fir zone in the Southern Rocky Mountains have been destroyed. *Photos by Michael Kuo*

Rocky Mountain Jacob's ladder features baseball-sized clusters of showy blue flowers and a foul, skunklike odor. *Photo by Carol Schmudde*

Wildlife in the spruce-fir zone is generally inconspicuous, and has adapted to the cooler temperatures and dampness that usually characterize the forests. Small groups of gray jays (*Perisoreus canadensis*) swoop quietly through the darkened branches in search of berries, but the birds are not bashful and will even accept food offered by humans. Red-breasted nuthatches (*Sitta canadensis*) scamper over tree trunks in search of food tucked into the bark, often appearing upside down, with their heads toward the ground. Solitary snowshoe hares (*Lepus americanus*), being primarily nocturnal, are almost never seen—but their scat and busywork (they often strip the bark from young twigs, saplings, and woody plants) are frequently encountered. Yellow-bellied marmots (*Marmota flaviventris*) inhabit open meadows at the edges of subalpine forests and make their home among rock piles and glacial moraines.

Red squirrels (*Tamiasciurus hudsonicus*) dart across branches, harvesting spruce cones and piling them in middens; the spruce seeds will be consumed later. Red squirrels are also avid harvesters of mushrooms (especially *Russula* species), which they dry on sunlit branches, storing the morsels for later consumption.

Black bears (*Ursus americanus*) forage for food throughout the summer and fall in the dense spruce-fir forests. Evidence of their presence might be claw marks on

Red columbines (Southern Rockies) and fairy slipper orchids are easily recognizable plants of subalpine forests. *Photos by Cathy L. Cripps*

Gray jays are stocky, gray birds with long tails and white heads that feature a black area near the nape. *Photo by Carol Schmudde*

Yellow-bellied marmots have an orangish underside and are found throughout the Rocky Mountains. *Photo by Michael Kuo*

trees or logs ripped open on the forest floor. Ants are a favorite food of black bears; and they claw apart downed logs to access the ants and their brood (eggs, larvae, and pupae). The high elevation offers them huckleberries, blackberries, and strawberries and, like the squirrels, bears will occasionally eat mushrooms. They often head to the lodgepole pine forests to dig for the underground "pogies," species of *Rhizopogon*, found with pines.

The mushrooms of the Rocky Mountain spruce-fir forests are numerous and often charismatic. During late summer monsoons, *Amanita muscaria* var. *flavivolvata*, beautiful boletes, and impressive species of *Cortinarius* creep out from the forests into the edges of meadows and clearings, flirting with sunlight. Golden chanterelle patches can produce mushrooms late into October at these elevations. If you're willing to venture into the dark tangle of trees, you will find litter-decomposing mushrooms like *Clavariadelphus* and *Infundibulicybe (Clitocybe) squamulosa*, arising from the dense mat of brown needles. When the subalpine forests "bloom" with mushrooms during a wet fall, the fungal diversity can be truly astounding.

Russula being dried by a squirrel. *Photo by Cathy L. Cripps*

Black bear and logs ripped up by bears looking for ants to eat in a subalpine forest. *Photo by Andy Hogg, Cathy L. Cripps*

Boletus rubriceps occurs in the Southern Rockies while *Boletus edulis* is the comparable species in the Northern Rockies. *Photos by Michael Kuo, Cathy L. Cripps*

Photo by Michael Kuo

Amanita muscaria var. *flavivolvata* (Singer) D. T. Jenkins

DESCRIPTION: CAP 5–25 cm across; nearly oval or round at first, becoming convex, then broadly convex to flat; sticky when fresh; adorned with numerous small, cottony warts that are initially yellow but very quickly fade to white; surface otherwise smooth, deep to bright red (but sometimes fading with age to pale orange or yellow); the margin sometimes becoming slightly lined. **GILLS** free from the stalk, broad in center, close or crowded; white. **STALK** 5–18 cm long; up to 3 cm thick; more or less equal above a swollen base; smooth at top and shaggy in lower half; white; with a high, skirtlike ring, with concentric bands of universal veil material at the top of the bulb and/or on the lower stalk. **FLESH** white; unchanging when sliced; odor not distinctive. **SPORE PRINT** white.

ECOLOGY: Mycorrhizal with a variety of conifers, growing alone, scattered, or gregariously; summer and fall; this bright red variety is distributed throughout the Southern Rockies but absent in Montana and northern Wyoming, where it is replaced by a yellow-capped *Amanita muscaria* (Pg 92) that has been called variety *formosa* and might be variety *guessowii.* The yellow variety at least also occurs with aspen.

OBSERVATIONS: The Rocky Mountain version of this classic, storybook "toadstool" features a yellow universal veil that soon fades to white. It is often found alongside *Boletus rubriceps* (Pg 179). All varieties of *Amanita muscaria* are dangerously poisonous and should not be eaten.

Photo by Cathy L. Cripps

Amanita pantherina (de Candolle) Krombholz (pale form)

DESCRIPTION: CAP 4–12 cm across, convex, shallow convex, becoming almost flat; smooth, slightly sticky; pale buff to light brown, yellow-brown; with scattered pieces of white tissue (warts) over the surface; margin with faint lines. **GILLS** free, crowded, broad in center, white; in young, covered with a white partial veil. **STALK** 4–14 × 2–3 cm, gradually larger toward the base down to a rounded bulb that is covered with a cup (volva) that has a collar at the top; dingy white; rather smooth; ring white, membranous, hanging skirtlike on upper part of stalk. **FLESH** white; odor not distinctive. **SPORE PRINT** white.

ECOLOGY: Scattered on soil in conifer and mixed conifer-aspen forests in the Rocky Mountains and mycorrhizal with both conifers and aspen; fruiting in summer and early fall.

OBSERVATIONS: *pantherina* for the spots of a panther. The typical *A. pantherina* is darker brown and is widely distributed in the West and elsewhere. The pale Rocky Mountain form may be a distinct taxon; another pale variety often called *multisquamosa*, appears in ponderosa pine forests. All can cause serious poisonings and should be avoided.

Photo by Michael Kuo

Hygrophorus olivaceoalbus (Fries) Fries

DESCRIPTION: CAP 3–12 cm across; convex, becoming broadly convex or more or less flat; slimy; with a streaked appearance resulting from stretched-out fibrils beneath the slime; dark brown to gray-brown, lighter toward the margin. **GILLS** attached to the stalk or running down it; close or nearly distant; white; waxy. **STALK** 3–15 cm long; up to 3 cm thick; equal or with a slightly swollen base; when fresh sheathed with slime over the lower portion; white at the apex; covered beneath the slime with brown fibrils that stretch out as the mushroom matures and often develop vaguely concentric bands or stripes; sometimes with a fragile and somewhat gelatinized ring. **FLESH** white; unchanging when sliced; odor and taste not distinctive. **SPORE PRINT** white.

ECOLOGY: Mycorrhizal with Engelmann spruce; growing scattered, gregariously, or in small clusters; late summer and fall; widely distributed in the Rocky Mountains.

OBSERVATIONS: *Hygrophorus olivaceoalbus* belongs to a fascinating group of "waxy caps" that feature not one but two partial veils protecting the young gills. The inner veil is composed of tiny threadlike fibrils, and the outer veil is composed of thick slime. When the mushroom approaches maturity, the cap expands to break the veils, which then are left to form a sheath around the stalk.

Photo by Michael Kuo

Hygrophorus pudorinus (Fries) Fries

DESCRIPTION: CAP 3–12 cm across; convex, becoming broadly convex or somewhat bell-shaped; slimy when fresh; bald or, with age, developing fine scales and cracks; pale pinkish orange or pinkish tan; the margin at first inrolled and cottony. **GILLS** attached to the stalk or beginning to run down it; close or nearly distant; white, unchanging or becoming yellowish to pinkish with age; waxy. **STALK** 4–9 cm long; up to 3 cm thick; more or less equal above a tapered base; dry; finely dotted with whitish tufts of fibrils near the apex that turn reddish brown when specimens are dried; whitish; often turning yellowish where handled or near the base. **FLESH** firm; white, or slightly pinkish to yellowish or orangish in the stalk base; unchanging when sliced, or turning slightly yellowish; odor often soapy and fragrant, or slightly unpleasant; taste not distinctive. **SPORE PRINT** white.

ECOLOGY: Mycorrhizal with spruces and other conifers; growing scattered to gregariously; late summer and fall; widely distributed in the Rocky Mountains.

OBSERVATIONS: This stocky species of *Hygrophorus* appears almost like a *Tricholoma*. A fragrant, orange-bruising variety of *H. pudorinus* with an orangish yellow stalk base (inside and out) is also found in the Rockies. Inedible.

Photo by Michael Kuo

Lactarius deliciosus var. *areolatus* A. H. Smith

DESCRIPTION: CAP 3–15 cm across; convex, becoming broadly convex, flat, or shallowly vase-shaped; slightly sticky when very fresh but soon dry; the margin sometimes inrolled when young; bald or slightly roughened, becoming rugged and subscaly with age when exposed to sun and dry conditions; carrot-orange or orange when fresh and when growing in the shade, but soon dull brownish yellow or dirty buff when exposed; with or without concentric zones of color; becoming stained and discolored greenish by maturity. **GILLS** attached to the stalk or beginning to run down it; close; orange, bruising very slowly red, then green; frequently entirely greenish at maturity. **STALK** 2–6 cm long; up to 3 cm thick; more or less equal; bald; without potholes; often becoming hollow; orange overall; bruising and staining like the cap. **MILK** very scanty (often absent); orange; staining tissues slowly deep red, then green. **FLESH** whitish to pale orangish; staining slowly deep purplish red when sliced, especially in the stalk base and over the gills; odor not distinctive; taste mild or slowly slightly acrid. **SPORE PRINT** pale yellowish.

ECOLOGY: Mycorrhizal with conifers; growing alone, scattered, or gregariously; late summer and fall; widely distributed in the Rocky Mountains.

OBSERVATIONS: This is the western North American version of the well-known European species *Lactarius deliciosus*. Despite its name, it is not particularly delicious. The name *areolatus* was originally meant to represent the areolate (cracked and scaly) cap—but specimens commonly encountered that are not exposed to direct sun or dry conditions have bald to merely slightly roughened caps.

Photo by Michael Kuo

Lepiota clypeolaria (Bulliard) P. Kummer

DESCRIPTION: CAP 2–8 cm across; bell-shaped, becoming broadly bell-shaped or nearly flat with age; dry; hairy to scaly; brown to brownish or yellow-brown; evenly colored (the center not usually contrasting with the rest of the surface). **GILLS** free from the stalk; close; white. **STALK** 4–10 cm long; usually under 1 cm thick; more or less equal; hairy to shaggy below; whitish, sometimes discoloring yellowish brown with age or on handling; with a sheathing white ring or ring zone that sometimes disappears. **FLESH** white; not changing when sliced; odor not distinctive. **SPORE PRINT** white.

ECOLOGY: Saprobic on conifer litter; growing scattered or gregariously; summer and fall; widely distributed in the Rocky Mountains.

OBSERVATIONS: This species can be recognized by its dull brownish colors, the free white gills, and the sheathing ring or ring zone. A similar species, *Lepiota magnispora*, features somewhat brighter colors and a contrasting cap center, and differs microscopically; it also occurs in the Rockies. Several of the small Lepiotas are suspected of producing dangerous toxins, although the toxicity of this species is unknown.

Photo by Vera S. Evenson

Mycena pura (Persoon) P. Kummer

DESCRIPTION: CAP 2–6 cm across; convex or bell-shaped, becoming flattened; moist or dry; bald; the margin lined; lilac to purple or brownish purple when young, but often fading. **GILLS** attached to the stalk by a tooth; close or nearly distant; whitish, or sometimes slightly pinkish to purplish. **STALK** 4–10 cm long; 2–6 mm thick; equal; hollow; bald or with a few tiny hairs; whitish or flushed with the cap color. **FLESH** insubstantial; watery grayish to whitish; odor and taste radishlike. **SPORE PRINT** white.

ECOLOGY: Saprobic on terrestrial conifer litter and sometimes on hardwood debris; growing alone, scattered, or gregariously; summer and fall; widely distributed in the Rocky Mountains.

OBSERVATIONS: The small size, radishlike odor, and purple colors make members of this species complex fairly easy to recognize—although faded, nearly whitish specimens can be baffling. It can be confused with the usually larger edible *Lepista nuda* (also called blewits), which has a fragrant odor and pink spore print, and inedible lavender *Cortinarius* species, which have a rusty brown spore print and a cobwebby veil. Toxic, contains muscarine.

Photo by Michael Kuo

Russula cinereovinosa Fatto

DESCRIPTION: CAP 3–9 cm across; convex, becoming broadly convex to flat, with or without a shallow depression; sticky when fresh, but soon dry; bald and shiny; deep grayish purple to deep brownish purple; the margin often lined by maturity; the skin peeling roughly one-third to halfway to the center. **GILLS** attached to the stalk or running slightly down it; close; cream at first, but soon orangish yellow. **STALK** 3–6 cm long; up to 2.5 cm thick; more or less equal; bald; whitish, with an occasional purplish streak. **FLESH** white; brittle; not changing when sliced; odor not distinctive; taste slowly to moderately slightly acrid. **SPORE PRINT** dull yellow.

ECOLOGY: Mycorrhizal with conifers; growing alone, scattered, or gregariously; late summer and fall; southwestern Rocky Mountains.

OBSERVATIONS: This gorgeous southwestern species is very similar to a purple, conifer-loving West Coast species that is frequently mislabeled "*Russula atroviolacea*," but the original description for that species calls for a red-capped *Russula* associated with willows. *Russula cinereovinosa* was described from New Mexico in association with conifers.

Photo by Michael Kuo

Tricholoma equestre (Linnaeus) P. Kummer

DESCRIPTION: CAP 2–3.5 cm across; broadly convex, with a central bump; sticky when fresh; bald, or with a few appressed, radially arranged fibrils over the center; bright lemon-yellow, darkening to brownish yellow from the center outward; the margin not lined, somewhat incurved when young. **GILLS** notched or nearly free from the stalk; close; with frequent short gills; pale to bright yellow. **STALK** up to 4 cm long and 1 cm thick; equal, or slightly enlarged at the base; very finely hairy; pale yellow. **FLESH** white; unchanging when sliced; taste and odor mealy. **SPORE PRINT** white.

ECOLOGY: Mycorrhizal with Engelmann spruce and probably with other spruces and firs; growing alone, scattered, or gregariously in late summer and fall, in monsoon season; distributed throughout the Rockies.

OBSERVATIONS: Described here is the diminutive form of *Tricholoma equestre* found in the spruce-fir zone. A larger, stockier version is found in the Rockies under lodgepole pine; it is not known whether these versions represent the same species—or whether either version represents the same species as the original European species. North American versions have sometimes been labeled *Tricholoma flavovirens*, and were assumed, along with the European species, to be edible; however *Tricholoma equestre* contains an unknown toxin that when consumed in great quantity can cause a serious poisoning known as rhabdomyolysis.

Photo by Michael Kuo

Tricholoma saponaceum (Fries) P. Kummer

DESCRIPTION: CAP 3–8 cm across; conical-convex when young, expanding to broadly convex with a shallow central bump; tacky when fresh; bald; dark olive-green at first, becoming lighter yellowish green; the margin paler, often slightly inrolled. **GILLS** notched or nearly free from the stalk; close or nearly distant; with frequent short gills; whitish with hints of yellow. **STALK** up to 5 cm long and 3 cm thick; equal, or slightly enlarged or tapering at the base; bald; often with a substantial underground portion; whitish toward the apex but flushed with yellowish green elsewhere; bruising orangish brown, and at the base bruising pinkish to pink. **FLESH** white overall but pinkish orange in the stalk base; taste and odor unpleasant. **SPORE PRINT** white.

ECOLOGY: Mycorrhizal with conifers; growing alone, scattered, or gregariously in late summer and fall; distributed throughout the Rockies.

OBSERVATIONS: We have described and illustrated the greenish, Rocky Mountain version of the classic European species *Tricholoma saponaceum*. As for all the members in the group, distinctive features include the pinkish to orangish flesh in the stalk base, and the overall stature. A grayish brown version is also found in the Rockies; it is otherwise very similar. The soapy taste when cooked makes it inedible.

Photo by Cathy L. Cripps

Xeromphalina campanella (Batsch) Kühner and Maire

DESCRIPTION: CAP up to 3 cm across at maturity; convex, becoming broadly convex with a central depression and an arched margin; bald; dry; lined on the margin when wet; brownish yellow, rusty, yellowish, or dull orange; usually darker toward the center; fading. **GILLS** running down the stalk, distant or nearly so; with frequent cross-veins; pale yellow or orangish. **STALK** up to 5 cm long and 3 mm thick; equal; bald above, but finely hairy (orange-brown) at the base; wiry and tough; yellowish above, darker brownish red below. **FLESH** insubstantial; taste and odor not distinctive. **SPORE PRINT** white.

ECOLOGY: Saprobic on the deadwood of conifers; typically growing in dense clusters of many mushrooms on stumps or logs, but occasionally growing alone or gregariously; summer and fall; distributed throughout the Rockies.

OBSERVATIONS: This tiny mushroom typically fruits in astounding numbers; tightly packed clusters can cover an entire stump, making it appear orange from a distance. Hallmark features include the centrally depressed mature cap and the distant gills that run down the stalk. *Chrysomphalina chrysophylla* also grows in dense groups on conifer wood in the Rockies, but the gills are a golden yellow color. Also see *Xeromphalina cauticinalis* (Pg 81).

Photo by Michael Kuo

Agaricus amicosus Kerrigan

DESCRIPTION: CAP 7–18 cm across; convex, bulky at first, becoming broadly convex; dry; whitish underneath with brown, appressed fibrils that remain tightly affixed or become aggregated into small scales; bruising reddish. **GILLS** free from the stalk; close; pink when young, becoming dark chocolate brown to blackish; often bruising reddish. **STALK** 4–12 cm long; up to 3 cm thick; more or less equal above a slightly to moderately swollen base; whitish; bruising reddish orange; smooth; with a skirtlike, whitish to brownish ring; the base sheathed in whitish to brownish veil material. **FLESH** whitish; immediately changing to reddish or brownish red when sliced; odor not distinctive, or sometimes fruity; taste mild. **SPORE PRINT** dark chocolate brown.

ECOLOGY: Saprobic, decomposing conifer litter; growing scattered or gregariously; summer and fall; documented commonly in Colorado, reported from Wyoming, but probably distributed throughout the Rockies in high elevations.

OBSERVATIONS: The red staining and the brown fibrils on the cap and high-elevation location are key identifying features. There are numerous species of *Agaricus* in the Rockies, many of which are not well documented; and some produce serious GI upsets.

Photo by Cathy L. Cripps

Cortinarius anomalus (Fries) Fries

DESCRIPTION: CAP 5–7 cm across, shallow convex, slightly domed, smooth or roughened, moist; pale lavender but rusty tawny in center; margin bent down and remaining lavender. **GILLS** attached, narrow; pale lavender, then milk coffee color with lavender tones; cobwebby veil (cortina) lavender, leaving a ring zone on upper part of the stalk that becomes reddish brown with spores. **STALK** enlarging toward base; length 5–13 × 0.5–1.5 cm wide at apex to 2–2.5 cm at base; silky, whitish lavender, with herringbone pattern of yellow tissue in lower half. **FLESH** whitish, lavender in stipe; stipe hollow; odor radishlike. **SPORE PRINT** reddish brown; spores round.

ECOLOGY: Mycorrhizal; scattered in spruce-fir forests often in moss; fruiting in fall. Known from the Rocky Mountains, Pacific Northwest, and the West.

OBSERVATIONS: This beautiful stately fungus is one of many *Cortinarius* species with lavender tones, but this one has roundish spores; *C. caninus* is similar but is more brownish. *Cortinarius* species are prolific in Rocky Mountain spruce-fir forests in fall and are recognized by their (usually) red-brown spore color and cobwebby veil. This species is part of a group that needs to be sorted out, and European names may not be appropriate; not edible.

Photo by Michael Kuo

Cortinarius elegantio-montanus Garnica and Ammirati

DESCRIPTION: CAP 4–12 cm across; convex, becoming broadly convex or nearly flat; sticky when fresh; bald or sometimes finely silky in places; occasionally with a few small, brownish scales over the center; yellow when young, becoming brownish yellow or yellowish brown. **GILLS** attached to the stalk; close, moderately broad; dull yellow at first, becoming cinnamon to rusty brown, sometimes with an olive tint. **STALK** 4–7 cm long; up to 3 cm thick; equal above a large, rimmed basal bulb; yellow when young, maturing to dull yellow; often discoloring brownish; dry; often with a ring zone; universal veil yellow, leaving remnants on the edge of the bulb. **FLESH** whitish in cap and pale yellow in stalk; frequently turning pink in the stalk when sliced; odor not distinctive or sometimes foul. **SPORE PRINT** rusty brown.

ECOLOGY: Mycorrhizal with Engelmann spruce; July and August; growing alone, scattered, or gregariously; southwest Colorado north to at least Wyoming.

OBSERVATIONS: This recently named species is very common in the spruce-fir zone and can be recognized by its sticky, dull yellow cap and large, rimmed basal bulb. It lacks the lavender tints of morphologically similar species.

Photo by Lee Gillman

Boletus rubriceps D. Arora and J. L. Frank

DESCRIPTION: CAP 10–30 cm across at maturity; convex in the button stage, becoming broadly convex to nearly flat; greasy to tacky; bald or occasionally breaking up to form some small scales, or rugulose; brownish red to reddish brown; sometimes with a whitish bloom when young. **PORE SURFACE** whitish and "stuffed" at first, becoming yellowish to brownish yellow, and eventually olive; not bruising; pores 2–4 per mm at maturity; tubes to 3 cm deep. **STALK** 10–18 cm long; 3–8 cm thick; swollen and club-shaped when young, becoming club-shaped or equal; finely reticulate-netted over at least the upper portion; white or pale brownish. **FLESH** white; solid; unchanging when sliced, or changing slightly pinkish; odor not distinctive; taste nutty. **SPORE PRINT** olive; spores larger than those in the *Boletus edulis* group.

ECOLOGY: Mycorrhizal with Engelmann spruce and perhaps with other spruces and firs; growing alone, scattered, or gregariously in late summer and fall, in monsoon season; distributed throughout the Southern Rockies.

OBSERVATIONS: Easily recognized by its large size and its greasy, reddish cap, this deliciously edible species is often identified as *Boletus edulis*—but that name represents a group with brown caps. This newly named species fits the reddish capped forms in the Southern Rockies that look similar to the European *Boletus pinophilus*, which genetic studies distinguish as a separate species. We find this version of "porcini" to be especially good when sliced, dried, and reconstituted in sauces. In the Northern Rockies, a brown-capped *B. edulis*-like species (Pg 164) is common.

Photo by Michael Kuo

Albatrellus confluens (Albertini and Schweinitz) Kotlaba and Pouzar

DESCRIPTION: CAP 3–20 cm across; irregular in outline; loosely convex, flat, or irregular; often fused with other caps; dry; bald; sometimes becoming cracked with age; pale orange, pinkish buff, or orangish, fading to tan—or sometimes becoming green as a result of mold growth. **PORE SURFACE** descending the stalk; white to cream; sometimes staining faintly greenish or yellowish; 3–5 pores per mm; tubes to 5 mm deep. **STALK** 3–6 cm long; 1–3 cm wide; usually a little off-center; whitish, developing tan discolorations; bald. **FLESH** whitish; soft when fresh; odor not distinctive; taste mild, or slightly foul and cabbagelike. **SPORE PRINT** white.

ECOLOGY: Mycorrhizal with conifers; usually growing gregariously; summer and fall; distributed throughout the Rockies.

OBSERVATIONS: This mushroom can appear in large numbers, with caps fused together. It is a slow growing, long-lasting species, and in the right weather conditions fruiting bodies can last for many weeks, eventually becoming green from the mold *Cladosporium*. Barely edible and not choice, it can have an "off" flavor. *Albatrellus ovinus* is also common in the Rockies. It is similar to *A. confluens* but is white with yellow tinges.

Photo by Michael Kuo

Alloclavaria purpurea (Fries) Dentinger and D. J. McLaughlin

DESCRIPTION: 2.5–10 cm high; 2–6 mm thick; cylindrical to nearly spindle-shaped; unbranched; sometimes somewhat flattened or featuring a groove or a twist; dry; soft; dull purple to purplish brown; paler at the extreme base; usually with a bluntly pointed tip. **FLESH** whitish to purplish; thin; odor and taste not distinctive. **SPORE PRINT** white.

ECOLOGY: Traditionally presumed to be saprobic, but recent investigators (Dentinger and McLaughlin 2006) suggest the possibility that it is mycorrhizal with conifers or associated with mosses; usually growing in tightly packed clusters under conifers; summer and fall; distributed throughout the Rockies.

OBSERVATIONS: This funky mushroom, common in spruce-fir forests, looks like a group of wiggling, purple worms standing on their heads. When fresh and purple, the species is fairly unmistakable—but older, faded specimens can be more difficult to identify. Also called *Clavaria purpurpea*, although genetic work has found that it is in a separate genus. It is considered edible.

Photo by Vera S. Evenson

Clavariadelphus ligula (Shaeffer) Donk

DESCRIPTION: FRUITING BODIES slender club shapes, 3–8(10) cm long × 1–2 cm across and more or less flattened; bases tapered, slender and firmly attached to conifer debris and soil by sometimes copious white mycelium; surface pale yellow-ocher to tan, dull, glabrous, becoming coarsely wrinkled. **FLESH** whitish, punky in upper part, firm at base, not staining when cut; odor mild; taste rather bitter. **SPORES** in mass white.

ECOLOGY: Fruiting sometimes by the hundreds in shady conifer forests, often spruce, in the Rocky Mountain region. These little saprobes are widely distributed in the late summer to early fall.

OBSERVATIONS: *Clavariadelphus sachalinensis* is very similar but its spores are larger than *C. ligula* and are ocher in color. Since spores are forcibly discharged from the surfaces of these basidiomycetes, spore prints can be made from them by placing the fresh fruiting body on its side on a piece of white paper and keeping it moist under a bowl or in waxed paper. The Rocky Mountains also harbor *C. pistillaris* and *C. truncatus*; the latter is larger, has a truncated cap, and is considered an excellent edible.

Photo by Ed Barge

Hydnum repandum Linnaeus

DESCRIPTION: CAP 4–10 cm, shallow convex, often depressed in center; dry, smooth; pale orange; margin turned down or under. **TEETH** spore-bearing surface in the form of small spines or icicles that hang down, a few mm long; running down the stalk; white. **STALK** 2–6 × 1–2 cm, sometimes larger at base when joined with others; dry, smooth; whitish or pale orange. **FLESH** white; brittle, breaking easily; odor mushroomy; taste mild. **SPORE PRINT** white.

ECOLOGY: Mycorrhizal; often in dense groups in spruce-fir forests in late fall, from the end of August into October throughout the Rockies; also found throughout North America and in Europe.

OBSERVATIONS: also known as *Dentinum repandum*, this fungus is easily recognized by its pale orange cap, white teeth, and brittle flesh. At first glance, it can look like a chanterelle because of the orange color. It is one of the last choice edibles to fruit in fall. *Hydnum umbilicatum* is slimmer with a hole in the center and *Hydnum albidum* is whiter; all are edible. They are known locally as "sweet tooths" or "hedgehog" mushrooms.

Photo by Vera S. Evenson

Polyozellus multiplex (Underwood) Murrill

DESCRIPTION: FRUITING BODY a cluster of individuals joined together, violet blackish fan shapes with irregular incurved and lobed margins, up to 10 cm broad. **STALKS** dark purplish black; dry; brittle, fused, up to 5 cm long and 1–2 cm wide; spore-bearing surface covers the upper parts of the fused stalks with purple-violet low ridges and veins, often forking and at times seeming to be poroid. **FLESH** soft but breaking easily; violet; odor mild, taste mild. **SPORES** white in deposit.

ECOLOGY: Not common but known from many parts of coniferous forests in the Rocky Mountains, especially in the northern regions or at higher elevations; mycorrhizal with spruce; fruiting in summer and fall.

OBSERVATIONS: Known commonly as the blue chanterelle, this well-camouflaged fungus is edible if you can find it! Its clustered purple-violet fruiting bodies distinguish it from the golden chanterelle (Pg 137), but the veined and ridged spore-bearing surface (the hymenium) is similar in both of these interesting, but not closely related, fungi. *Gomphus clavatus* is flat, funnel-shaped, with a wrinkled purple hymenium, but it is much fleshier; it is also edible. *Polyozellus multiplex* is highly prized as a dye source.

Photo by Michael Kuo

Sarcodon imbricatus (Linnaeus) P. Karsten

DESCRIPTION: CAP 5–25 cm across; convex to broadly convex with a central depression (the depression is sometimes perforated with age); dry; conspicuously covered with coarse, raised, dark brown to blackish scales; pale to dark brown underneath the scales; the margin inrolled. **UNDERSURFACE** running down the stalk; covered with spines or "teeth" that are 0.5–1.5 cm long; pale brown at first, becoming darker with age. **STALK** 4–10 cm long; up to 3.5 cm thick; equal, or slightly enlarged at the base; bald, except where punctuated by aborted spines; whitish or pale brownish; base with white mycelium. **FLESH** white to pale brownish; unchanging when sliced; odor not distinctive; taste mild or bitter. **SPORE PRINT** brown.

ECOLOGY: Mycorrhizal with conifers such as spruce; growing alone, scattered, or gregariously in late summer and fall; distributed throughout the Rockies.

OBSERVATIONS: This is a very common, large, charismatic fungus; it can reach extremely large sizes. The spines on the undersurface of the cap are covered with spores—the result of an ingenious design to increase the spore-bearing surface area. It is a good, solid edible with a strong flavor, but older specimens can taste bitter, especially to some people. Known locally as "hawkwing" or "scaly urchin." The *Sarcodon scabrosus* complex is less scaly, more blackish gray, and bitter tasting, so caution is advised when picking for the table.

Photo by Michael Kuo

Sarcosphaera coronaria (Jacquin) J. Schröter

DESCRIPTION: FRUITING BODY bowl-shaped, ball-like, or goblet-shaped when young, with a small opening near the top; usually splitting into "rays" by maturity and folding back to be star-shaped or roughly saucer-shaped; up to 20 cm across when mature; inner surface whitish becoming lilac to pale lilac-brown; bald or finely scaly; outer surface whitish, roughened (and covered with dirt), sometimes bruising yellowish; stalk rudimentary or absent. **FLESH** whitish; brittle; odor not distinctive. **SPORE PRINT** white.

ECOLOGY: This member of the Ascomycota is mycorrhizal with conifers; usually growing in clusters, partially submerged in the ground; most often found in spring or early summer, but also fruiting in summer and fall; distributed throughout the Rockies.

OBSERVATIONS: When young, the clustered cups are nearly closed, and grow submerged in the ground with only the top portion sticking out. With age, the cup opens up and the edges split and peel backward in starlike rays.

Photo by Vera S. Evenson

Scutellinia scutellata (Linnaeus) Lambotte

DESCRIPTION: FRUITING BODIES in the form of stalkless cups, at first nearly spherical, spreading and flattening to discs with age; 0.5–2 cm across; exposed upper surface smooth, bright reddish orange to scarlet; margins turned up and distinctively decorated with long, sharply pointed, dark brown to blackish hairs, often referred to as eyelashes; underside of cups pale brownish orange, scattered with short dark hairs. **STALK** absent, attached directly to substrate. **FLESH** very thin, pale; odor mild. **SPORES** pale off-white.

ECOLOGY: These charismatic little saprobes grow directly attached to damp decayed wood and in moist soil among dead plant remains, often in subalpine forests and along montane streams; midsummer to early fall.

OBSERVATIONS: *scutellata* for the Latin word for a small shield. There are several yellowish orange to bright red members of the genus *Scutellinia* (eyelash cups) in the Rocky Mountains, especially in montane to subalpine ecosystems, always in moist habitats. They differ by length of the hairs and spore characteristics.

Photo by Cathy L. Cripps

SNOWBANKS

In spring, the Rocky Mountains are gradually transformed from snowy white to shades of green by warming temperatures. By late June and into July, only the higher forests, ridges, and peaks are still covered in snow. At subalpine elevations, the snow melts unevenly because of local differences in topography and microclimate. This interplay leaves a patchwork of *remnant snowbanks* that are alternately reduced by melting on warm days and refrozen into more persistent structures by cold nights.

In the open patches of earth around the snowbanks, sunshine-colored avalanche lilies and delicate white spring beauties bloom in abundance. Avalanche lilies, also known as glacier lilies (*Erythronium grandiflorum*) are recognized by their brilliant yellow color and six recurved petals; their prominent anthers are typically yellow but are red in some populations. The mature pods are browsed by deer, elk, and bighorn sheep. The lilies produce subterranean corms that have been used as food by humans, but harvesting kills the plant. The tiny white spring beauties (*Claytonia lanceolata*) also produce edible corms that are excavated by bears and rodents for a meal.

Near the snowbanks, black felty patches, observed on the lower branches of spruce trees, are a result of the conifer snow mold (*Neopeckia coulteri*) enveloping the needles. The mold proliferates below the snow surface and is revealed with spring thaw; it is an indicator of winter snow depth.

One of the few birds to overwinter at these high elevations, the dark-eyed junco (*Junco hyemalis*), can be heard twittering among the trees or spotted flashing its white tail feathers in flight. The other persistent resident is the Steller's jay (*Cyanocitta stelleri*) with its stunning blue body and black crest. These birds make their living as omnivores eating a variety of foods, including seeds and arthropods.

Avalanche lilies. *Photo by Ed Barge*

A fascinating aspect of this habitat is that life not only flourishes around, but in and under the snowbanks. The so-called "subnivean zone" is a cozy place for an array of creatures and microbes; the air space is warmed by the soil and insulated by the blanket of white snow. Rodents such as voles, mice, and ground squirrels tunnel through the snow; the extent of their tunneling is apparent when their subnivean nests and latrines are revealed with spring melt. The white weasels of winter, called ermine, frequent the snow tunnels as predators of rodents. Snowshoe hares will tunnel a few feet into the snow for protection and warmth.

Periodic outbreaks of snow fleas (*Hypogastrura nivicola*) or springtails dot snowbanks like animated black specks in warmer weather. Springtails can snap their

Black conifer snow mold on a young spruce indicates winter snow depth. *Photo by Cathy L. Cripps*

Steller's jay with its royal blue feathers and black crest. *YNP Photo*

A rodent tunnel melting out and golden-mantled ground squirrels playing in spring. *Photos by Cathy L. Cripps*

The tiny orange cups of the dung fungus (*Byssonectria cartilagineum*) on a melted-out rodent nest, its typical habitat. *Photos by Cathy L. Cripps*

bodies in such a way as to throw themselves through the air. In more sunlit areas, a pink color often tints the surface of the snowbanks. This phenomenon is known as "watermelon snow" because of its color and odor (take a whiff!). It is a result of the snow algae (*Chlamydomonas nivalis*) producing orange pigments (anthocyanins) in spring sunlight. However, toxins are produced by the algae, which should rule out any thought of making snow cones.

Deeper, at the snow-soil interface, vast microbial mats of fungi and bacteria have been discovered functioning at near freezing temperatures. Many are molds and microscopic fungi previously unknown to science. When sun and topography conspire to melt the snowbanks quickly on warm afternoons, the vast delicate networks of mycelium formed under the snow are revealed.

In subalpine forests of the Rocky Mountains and mountains to the west, the larger "snowbank fungi" show up as spring proceeds. Since the meltwater sustains fungal fruiting, an array of fruiting bodies can be observed popping up from beneath the snow, ringing snowbank edges, or standing as sentinels that mark where snowbanks once existed. The mushrooms produce their own melt-cavities that function like tiny caves of warmth when the sun angle is just right. This group of fungi was first noticed ringing snowbanks by several early mycologists (W. B. Cooke, O. K. Miller, and A. H. Smith) who named and recorded the various fungal forms. Today, we know that this snowbank mycoflora consists of a disparate group of gilled and nongilled mushrooms with ecological roles as saprobes, wood decomposers, mycorrhizal fungi, and even a few pathogens.

Yellow-veiled *Cortinarii* and blue, gray, and white *Hygrophorii* are the primary mycorrhizal species in the snowbank mycoflora. The pristine white and meaty *Hygrophorus subalpinus* is sometimes gathered for food. The silvery *Clitocybe glacialis* and frosty tan *C. albirhiza* are two of the most conspicuous terrestrial saprobes. Special wood decomposers such as the graceful *Mycena overholtsii* and the saw-tooth

Snowfleas (springtails) on a mushroom cap. *Photo by Cathy L. Cripps*

Mycelial networks formed under the snow revealed as snowbanks recede. *Photo by Cathy L. Cripps*

Mycena overholtsii fruiting in a snow-melt cavity. *Photo by Cathy L. Cripps*

gilled *Lentinellus montanus* form on logs, arising out of pockets of melting snow. Brilliant yellow lemon drops (*Guepiniopsis alpina*) and bright orange-blue cups (*Caloscypha fulgens*) dot the black-and-white landscape. The hefty false morel (*Gyromitra montana*) appears like a large brown brain arising without a body from under snowbank edges. While not recommended for eating because of its toxic cousins, it is a spectacular fungus.

Early spring in the Rockies with its combination of mud, slush, and snow does not seem like a time or place for mushrooming, but these special fungi are worth the trudge. Sometimes they can be found more conveniently along dirt roads where snow has receded. Although it is possible to find the "snowbankers" in lower-elevation lodgepole pine forests, they are more likely to be spotted in spruce-fir forests in late May through early July from Montana to Colorado. They also occur in whitebark pine forests. Certain snowbank species are known from other parts of the world, but it is only in western North America that they come into their own as a unique ecological group closely tied to melting snowbanks.

Orange cups of *Caloscyphe fulgens* and snowbank false morels fruiting at the edges of spring snowbanks. *Photos by Cathy L. Cripps*

Photo by Cathy L. Cripps

Clitocybe albirhiza H. E. Bigelow and A. H. Smith

DESCRIPTION: CAP 2–4 cm across; broadly convex, some with a slight depression and others with a slight bump in the center; smooth; mostly whitish tan, some with a pale pink tint, with a white frosty covering that wears off on weathering; margin turned down or under, sometimes with a white rim. **GILLS** attached or running slightly down the stalk; narrow, thin; cream, buff with age. **STALK** 2–4 × 0.5–1 cm, equal or narrower at the top or middle; whitish, cream, with a frosty coating; with copious white rhizomorphs at the base (dig it up!). **FLESH** a pale watery buff; odor flowery or floury. **SPORE PRINT** white.

ECOLOGY: A saprobic fungus that occurs in groups near melting snowbanks, or where snowbanks have melted out, often in protected areas under trees; typically in spruce-fir but also in lodgepole pine forests from late May to early July. This species is recorded from high-elevation forests in Colorado, Idaho, Montana, Utah, and Wyoming in the Rockies and also from the Pacific Northwest.

OBSERVATIONS: *albirhiza* for the copious white rhizoids at the base of the stalk. These rhizoids distinguish it from *C. glacialis* (Pg 194). *Clitocybe pruinosa*, which has a lead-gray cap, is another rhizomorph-former occurring near snowbanks in the Rockies. *C. albirhiza* and *C. glacialis* often occur together and they both age to a watery yellow-brown making them difficult to distinguish. Clitocybes are not for eating.

Photo by Cathy L. Cripps

Clitocybe glacialis Redhead, Ammirati, Norvell, and M. Seidl

DESCRIPTION: CAP 2–5 cm across, broadly convex with a turned-down margin; smooth, greasy or silky dry; silvery gray, with a hoary frosted look, more gray-brown with age. **GILLS** narrowly attached, thin, a bit crowded or not; pale gray to gray-brown. **STALK** 2–3.5 × 0.5–1.5 cm, equal; silvery pale gray with a hoary coating. **FLESH** watery gray; odor indistinct. **SPORE PRINT** white.

ECOLOGY: Fruiting at the edges of snowbanks or in cavities melted out of snowbanks, and persisting in areas where snowbanks previously existed; scattered in high-elevation spruce-fir, whitebark pine, and lodgepole pine forests in late May to early July and from Montana to Colorado. This fungus apparently decomposes recalcitrant organic matter in the soil.

OBSERVATIONS: *glacialis* for its cold-loving nature and association with snowbanks, but it was once called *Lyophyllum montanum* (A. H. Smith). As fruiting bodies weather, they turn a gray-brown and can be confused with *C. albirhiza*, which is distinguished by its white rhizoids at the base of the stalk. *Hygrophorus marzuolus* (Pg 195) is another gray snowbank mushroom, but it is larger with thicker well-separated gills. The flesh of these mushrooms remains cold after picking and on a hot spring day the mushrooms can be used to cool a hot forehead, but they are not edible.

Photo by Cathy L. Cripps

Hygrophorus marzuolus (Fries) Bresadola

DESCRIPTION: CAP 3–6 cm across; convex often with a low dome, becoming broadly convex; thick-fleshed; steel gray, sometimes silvery in the center, with white streaks; smooth, greasy; margin curled under. **GILLS** narrowly attached or running slightly down the stalk; thick, waxy, well-separated; whitish with a pale gray tone. **STALK** stout, 3–8 cm long × 1–3 cm wide, equal or gradually enlarged toward the base; smooth, greasy; whitish with gray tones, hollow in older fruiting bodies. **FLESH** white with water-soaked appearance; odor indistinct. **SPORE PRINT** white.

ECOLOGY: Occurring as scattered fruiting bodies near snowbanks at high elevations in the Rocky Mountains and other western mountains in spring; in spruce-fir, whitebark pine, and lodgepole pine forests and likely mycorrhizal with these trees.

OBSERVATIONS: *marzuolus* refers to a March fruiting, which is more likely in Europe than in the Rockies. This large silvery gray mushroom can be confused with *Clitocybe glacialis* (Pg 194), which is smaller with thinner gills, or with *Hygrophorus caeruleus* (O. K. Miller), which has a more bluish gray coloration and strong rancid odor. All occur in the same habitats, and while all are large and fleshy, none is recommended for eating.

Photo by Cathy L. Cripps

Hygrophorus subalpinus A. H. Smith

DESCRIPTION: CAP 5–10 cm across or larger, shallow convex, pure white, smooth, greasy, with margin turned under. **GILLS** attached or running slightly down the stalk; narrow, thick; white; at first covered with a glutinous ring. **STALK** robust, 4–6 cm long × 2.5–4 cm wide, with rounded bulbous base; white; smooth, sticky. **FLESH** white; firm, solid; odor indistinct. **SPORE PRINT** white.

ECOLOGY: Fruiting at the edges of snowbanks in spring in the Rocky Mountains and the Pacific Northwest; mycorrhizal and typically found half buried in duff in spruce-fir forests from late May to early July. This species appears to be endemic to North America.

OBSERVATIONS: *subalpinus* refers to its high-elevation habitat. This meaty species can occur in quantity and has been sold in markets in the Pacific Northwest. The firm texture of the flesh lends itself to cooking, but the flavor is bland or metallic when eaten in quantity. Similar-looking white mushrooms in this habitat include smaller white *Hygrophorus* species or the larger *H. gliocyclus* (Pg 213), which is cream-colored and does not fruit until early summer. As a caution, early spring whitish Amanitas (Pg 212) can look similar in the button stage.

Photo by Cathy L. Cripps

Lentinellus montanus O. K. Miller

DESCRIPTION: CAP 4–6 cm across or larger, shell-shaped, sometimes with a wavy margin; dark brown in the center and pale brown toward the margin; dry; slightly bumpy or rough hairy. **GILLS** radiate from an attachment point on one side, and are interspersed with shorter gills; well-separated; cream or buff; edges strongly saw-toothed when mature. **STALK** absent. **FLESH** thick, rubbery; cream; odor mushroomlike; taste mild. **SPORE PRINT** white; spores with short spines, amyloid.

ECOLOGY: A saprobe typically on conifer wood surrounded by remnant snowbanks, but also persisting after the snow has melted; occurring in spruce-fir forests on logs and branches of downed trees from May to July. Reported from Montana and Idaho, south to Colorado and Utah, and from other areas of the West.

OBSERVATIONS: *montanus* for its mountain habitat. Orson Miller first described this species and reports it as edible but tough. Its shelflike form looks superficially like an oyster mushroom, but it is tougher and has toothed gills. *Neolentinus ponderosus* (Pg 79) is another tough fungus with serrated gills that is larger, stalked, and does not fruit until later in the year. *Lentinellus montanus* is related to nongilled fungi such as *Auriscalpium vulgare* (Pg 84), the "toothpick" fungus.

Photo by Cathy L. Cripps

Mycena overholtsii A. H. Smith and Solheim

DESCRIPTION: CAP 2–4 cm across; bell-shaped, flatter when opened; smooth, greasy; pale gray, gray-brown or darker gray, often with a yellow-brown center; margin faintly lined, sometimes with a white rim. **GILLS** attached or running down the stalk a bit, well-spaced; bright white; edges can be gray. **STALK** long, thin, rooting, 5–10 cm × 3–7 mm; equal, sometimes a bit curved; smooth in top half and fuzzy with white mycelium in the lower half; white or tinted pale orange. **FLESH** white or yellow-gray in lower stalk with a pearly luster; odor slight. **SPORE PRINT** white.

ECOLOGY: Occurring in clusters on wood near snowbanks, often in pockets where the snow has melted to form a cavity; this wood decomposer fruits in spruce-fir and lodgepole pine forests in early spring from May to July; known in the Rockies from Colorado to Alberta, and also from western coastal mountain ranges.

OBSERVATIONS: *overholtsii* for the North American mycologist, L. O. Overholts, who visited this habitat in Colorado in the 1920s. This graceful fungus is easily recognized as one of the few gilled mushrooms found in clusters on wood near snowbanks; the dense mycelium on the lower stalk is diagnostic and it has been called the "fuzzy foot" for this reason. This fungus has recently been found near high-elevation snowbanks in Japan.

Photo by Cathy L. Cripps

Neohygrophorus angelesianus (A. H. Smith and Hesler) Singer

DESCRIPTION: CAP 1–3 cm across; funnel-shaped, depressed in center; smooth, greasy; dark or milk chocolate brown but drying lighter, sometimes covered with a whitish bloom; margin turned down. **GILLS** go down the stalk but there is a definite line where gills meet the stalk; thick, waxy, well-separated, violaceous brown or wine color. **STALK** 2–4 × 0.2–0.4 cm, equal, a bit curved, smooth; same color as cap, sometimes with a whitish bloom; with white mycelium at base. **FLESH** brownish, but dries lighter; rubbery, turning red in KOH (test a wedge of cap and gill); odor absent. **SPORE PRINT** white, spores amyloid.

ECOLOGY: Rare or often unrecognized, fruiting near melting snowbanks in spring from May to early July; reported from Idaho, Montana, and Utah in the Rocky Mountains and also from the Pacific Northwest. Its ecological status is not entirely known but it usually occurs under spruce-fir or lodgepole pine.

OBSERVATIONS: This small fungus looks like a cross between a *Hygrophorus* and a *Clitocybe*, meaning that it has the waxy looking gills of the former and the coloration of the latter. However, it is distinctive because of the violaceous color of the gills and its habitat near melting snowbanks; the chemical test with KOH can help confirm the identification when possible.

Photo by Ed Barge

Cortinarius colymbadinus Fries

DESCRIPTION: CAP 4–6 cm across, convex with a dome or low bump in center; smooth, greasy; brown, orange-brown, olive-brown, drying lighter; margin covered with yellow tissue when young. **GILLS** attached, pale yellowish orange becoming orange-brown; covered with thin yellow cobwebby veil at first. **STALK** 4–6 × 1–1.5 cm, equal or slightly larger at the base; smooth; watery whitish pale brown; with lemon-yellow veil covering the lower part or only as bits of yellow tissue on stipe. **FLESH** watery pale brown; odor not distinctive. **SPORE** print brown.

ECOLOGY: Mycorrhizal, occurring in high-elevation conifer forests near melting snowbanks in early spring in the Rocky Mountains and the Pacific Northwest; typically in spruce-fir, whitebark pine, or lodgepole pine forests. Uncommon.

OBSERVATIONS: There are a number of yellow veiled *Cortinarius* species found near melting snowbanks, and this group needs further study. Some species can be delineated by their colors under fluorescent light: the veil of *C. ahsii* and the flesh of *C. colymbadinus* fluoresce yellow; the *C. flavobasilis* group fluoresces orange at the base of the stalk.

Photo by Cathy L. Cripps

Guepiniopsis alpina (Tracy and Earle) Brasfield

DESCRIPTION: FRUITING BODIES fat dished cups, 0.5–2.5 cm wide, 2 cm high, narrowing to a point of attachment; some with a tiny 1 mm stalk; interior of cup golden yellow, smooth, somewhat translucent; exterior also bright yellow but can be more red-brown at the base. **FLESH** with a jellylike consistency, but tougher toward the base; odor and taste indistinct. **SPORES** white.

ECOLOGY: Hanging in groups on logs and small to large branches that melt out of snow in spring; a saprobe, typically found on wood without bark. It is one of the first snowbank fungi to fruit in spring (May), but it can persist into summer. It shrivels in dry periods and then revives after rain and produces spores again.

OBSERVATIONS: *alpina* for its high-elevation habitat; also known as *Heterotextus alpinus*. Lab tests have shown that some strains are truly psychrophilic (cold-loving) because they grow better at colder temperatures. Other yellow, spring, cup fungi, members of the Ascomycota, do not have the jellylike consistency and swollen appearance of these "lemon drops."

Photo by Cathy L. Cripps

Pycnoporellus alboluteus (Ellis and Everhart) Kotlaba and Pouzar

DESCRIPTION: FRUITING BODY 5–30 cm, shelflike or flat against dead, downed wood; upper surface slightly textured, sometimes dimpled, dull to bright orange; margin ragged. **LOWER SURFACE** appearing toothed, of long tubes with ragged, eroded-looking mouths; white to orange, or mottled; tubular layer can be several cm thick. **STALK** absent. **FLESH** pale orange; spongy, marshmallow soft or papery; odor not distinctive. **SPORES** white.

ECOLOGY: As single or multiple fruiting bodies on downed, dead conifer wood without bark, near melting snowbanks in early spring from late May through June; dried fruiting bodies can persist into July or even August; a decomposer of conifer wood in subalpine habitats at high elevations.

OBSERVATIONS: *alboluteus* for the white to yellowish colors. There are two main snowbank polypores that have a marshmallow texture, this one and *Tyromyces leucospongia* (Pg 203), which is white; both can dry to a papery texture and persist for a long time. The polypore *Pycnoporus cinnabarinus* (Pg 46) is also bright orange, but it is tougher, has small pores, and is not found in snowbank habitats. *P. alboluteus* is possibly toxic, because of one report of a poisoning from Cameron Pass in Colorado.

Photo by Cathy L. Cripps

Tyromyces leucospongia (Cooke and Harkness) Bondartsev and Singer

DESCRIPTION: FRUITING BODY shelf-shaped, 3–8 cm or larger × 3–6 cm or wider; upper surface cottony, whitish or pale pinkish buff; margin soft and often folded over. **PORES** tiny (2–3 per mm), round with jagged mouths; with some layering; white but staining watery pale red-brown on handling or drying. **FLESH** soft like a marshmallow, cottony, or papery-dry; white; odor not distinctive. **SPORES** white.

ECOLOGY: At high elevations on downed conifer logs that have lost their bark; fruiting just as the snow melts from June to early July, but it can persist until later in the year. It produces a brown cubical rot of spruce and fir and also of whitebark pine and is known from most western states.

OBSERVATIONS: *leucospongia* for its white soft texture; however, this fungus can turn papery in dry weather; it is also called *Oligoporus leucospongia.* It co-occurs with the other snowbank polypore *Pycnoporellus alboluteus* (Pg 202), which is larger, has a pale orange color, and is more toothed underneath. Both are annuals that form each year and decompose cellulose in conifer wood.

Photo by Cathy L. Cripps

Caloscypha fulgens (Persoon) Boudier

DESCRIPTION: FRUITING BODY deeply cup-shaped with pulled-in margin, 2–4 cm across, with bright orange smooth interior; exterior dull orange, smooth, staining dark bluish green; with a knot of rhizomorphs at the base. **FLESH** brittle, thin, often cracking; odor not distinct. **SPORES** white.

ECOLOGY: Scattered on the ground near melting snowbanks often on mats of conifer needles at high subalpine elevations. These cups can appear in great number in early spring from late May through June in Colorado, Montana, Utah, Idaho, and Wyoming. It is a seed pathogen with the potential to kill spruce seeds in cold, wet soils such as those below snowbanks.

OBSERVATIONS: *fulgens* for the shiny appearance. This fungus is easily recognized by its colors; handling enhances the bluish bruising/staining reaction. Albino forms that lack carotenoid pigments are known, but the cups still stain bluish. While these cups are known in Europe, fruiting between March and May, it is only in western North America that they are considered part of the snowbank community. The orange *Aleuria aurantia* is larger, more irregular, does not stain bluish, and fruits in summer and fall in different habitats.

Photo by Cathy L. Cripps

Gyromitra montana Harmaja

DESCRIPTION: CAP 5–13 cm wide × 5–9 cm high or larger; a convoluted mass of wrinkles reminiscent of a brain or walnut; yellow-brown; with edges pressed to the stalk; underside of cap light-colored, whitish to buff. **STALK** can be massive, 2–8 cm high × 3–8 cm wide or larger, whitish with numerous chambers when cut open. **FLESH** brittle; white; odor not distinctive. **SPORES** white.

ECOLOGY: In small to large groups next to melting snowbanks in spring; fruiting from June to July, but occasionally found later at high elevations. Isotope signatures suggest it is a decomposer but it is always associated with conifers, particularly spruce and fir in the Rocky Mountains.

OBSERVATIONS: Once called *Gyromitra gigas*, which is a more fitting name because huge specimens do occur, it has been considered edible but is not recommended because its look-alike relatives contain gyromitrin. Although there are reports of unpleasant reactions from *G. montana* in the Rockies, the actual culprit is often hard to pin down. *Gyromitra esculenta* (Pg 140) is a seriously toxic look-alike that occurs in spring in the Rockies, and fruiting periods can overlap, making mistaken identity a problem.

Photo by Ed Barge

Plectania nannfeldtii Korf

DESCRIPTION: FRUITING BODIES cup-shaped, 1–1.5 cm wide, with black, smooth interior and upturned rim; exterior also blackish, with a frosted appearance. **STALK** generally long, up to 4 cm and a few mm wide; blackish, with blackish hairs, also appearing frosted; **FLESH** blackish-gray; odor not distinct. **SPORES** light-colored.

ECOLOGY: Occurring in groups at the edge of melting snowbanks in spring in the Rocky Mountains. This decomposer often fruits on mats of spruce and fir needles and can be found attached to small twigs of these tree species. It occurs in the Rockies from Colorado to Montana and in mountain ranges to the west from late May into July.

OBSERVATIONS: *nannfeldtii* for a Swedish mycologist. These cups are very difficult to see against the dark background of matted needles, but careful searching around snowbank edges at the right time is likely to turn them up. Walking slowly and bent over is the best technique for finding them. They can be confused with other black cup fungi, most likely *Plectania milleri*, which does not have a stalk but also can fruit in the spring.

Photo by Lee Gillman

HIGH-ELEVATION PINE FORESTS

The high-elevation pines of the Rocky Mountains are some of the most magnificent, ancient, and picturesque pines in the world. At the upper limits of the subalpine ecosystem, the pines exist as scattered flagships with gnarled branches reaching leeward on high ridges. They can form extensive pure forests of astounding stark beauty or a tree-line boundary of dense stunted trunks. In these harsh, dry sites, at elevations not reached by other forest types, the pines hold soil on steep slopes and retard snow melt, which helps to sustain stream flow throughout the summer. In the Southern Rockies, Rocky Mountain bristlecone pine (*Pinus aristata*) dominates the high-elevation pine forests and, in the north, whitebark pine (*Pinus albicaulis*) takes over this role; their ranges do not overlap. Limber pine (*Pinus flexilis*) spans the whole distance, joining bristlecone pine in the south and whitebark pine in the north, but it comes into its own as a forest tree in the Middle Rockies where

Ancient bristlecone pine in Colorado. *Photo by Lee Gillman*

Nutcrackers extract whitebark pine seeds from cones with long beaks. *Photo by Shawn T. McKinney*

the other pines are absent. Western white pine (*Pinus monticola*) skirts the western edges of the Northern Rockies. All are 5-needle pines, which makes them special inhabitants of the western states.

Ancient bristlecones can reach 2,500 years of age (back to 500 B.C.), and the oldest whitebark is recorded at 1,270 years. More typical might be 300–1,500 years for bristlecone and 150–500 years for whitebark pine. In Colorado, bristlecone pines grow at altitudes up to 12,000 feet near the tree-line and in Montana whitebark pine forms extensive forests at 9,000 feet. Limber pine spans the whole elevation range from the tree line down to the plains. Like whitebark pine, mature limber pines can reach 1,000 years of age.

"Ghost forests" composed of the gray twisted trunks of dead pines can be observed silhouetted against the skyline in some areas. These "skeletons" are often the result of a vulnerability to white pine blister rust (*Cronartium ribicola*) and mountain pine beetle (*Dendroctonus ponderosae*) infestations; these threats have increased in recent years.

Squirrels store the seeds for winter and grizzly bears raid the squirrel caches. *Photo by Andy Hogg*

The fat seeds of all 5-needle pines are important to wildlife such as squirrels, birds, and bears as a food source. Whitebark pine seeds lack the "wings" of normal pine seeds and the species is almost totally dependent on Clark's nutcrackers (*Nucifraga*

Bristlecone pinecones have spines and are covered in sap; whitebark pinecones are round and purplish; limber pinecones are long, curved and green or brown. *Photos by Lee Gillman, Cathy L. Cripps*

columbiana) to disperse its nuts! The birds pluck seeds out of cones and bury them in caches a distance away. Seeds not retrieved by the birds germinate and grow into trees. Bristlecone and limber pines are less reliant on the birds since their seeds have wings and also can be dispersed by wind. In whitebark pine forests, red squirrels (*Tamiasciurus hudsonicus*) compete for seeds by clipping and dropping whole branches to keep the cones from the birds. When whitebark pinecones mature in the fall, the whole forest can erupt in a flurry of activity with birds squawking and pecking on cones in the tops of trees and swooping through the forest, while squirrels chatter, clip, and gather cones from fallen branches. Squirrels cache whole cones in a pile for winter, but cagey grizzly bears (*Ursus arctos horribilis*) bide their time, and in late fall and early winter they raid the stashes for the meaty seeds.

All of these 5-needle pines are easily recognized by their high-elevation habitats, relatively light-colored bark, and needles in bundles of five. All have a similar craggy appearance but can be distinguished from each other by their cones and their geography. Bristlecone pines have cones with a bristle on each scale, and the needles, which are crowded on branches, are sprinkled with white resin flecks. Whitebark pines have round, purplish cones that do not open. Limber pinecones are long and green, and eventually turn brown as they open. In northern forests, if cones are observed under a tree, it is usually a limber pine because the round purple cones of whitebark are pretty much all eaten by birds and squirrels. Whitebark and limber pine trees are often composed of several trunks joined at the base. This is the result of several seeds in a bird cache germinating and individual seedlings uniting into one tree.

The forest floor beneath 5-needle pines is often barren, but understory plants include grouse whortleberry (*Vaccinium scoparium*) and, in the north, beargrass (*Xerophyllum tenax*). Wild gooseberry (*Ribes* spp.) is the known alternate host of white pine

Beargrass is a common understory plant at the northern extent of whitebark pine's range. *Photo by Peter Lesica*

Paintbrush can be an alternate host of white pine blister rust fungus. *Photo by Cathy L. Cripps*

blister rust (*Cronartium ribicola*) (Pg 219), but recently species of paintbrush (*Castilleja*) and lousewort (*Pedicularis*) also have been found to be reservoirs of the rust.

Bristlecone pine occurs in Colorado in the Sangre de Cristo Mountains, Spanish Peaks, Pike's Peak area, Front Range, San Juan Mountains north of Creede, and the Sawatch Range south of Rocky Mountain National Park. Whitebark pine occurs at most of the timberlines in Wyoming, including in the Gros Ventre, Tetons, Bighorns, and the Absaroka-Beartooth Mountains (the Wind Rivers are too dry). In Yellowstone Park, whitebark pine can be observed on Mount Washburn and Dunraven Pass. Northward, whitebark pine forests are scattered through the Sawtooths, Beaverhead, Madison, and Gravelly Ranges of Montana to Glacier National Park and into Alberta

Mushroom collecting habitat under whitebark pine. *Photo by Cathy L. Cripps*

Suillus ("slippery jack") fruiting bodies chewed on by an unknown mammal. *Photo by Cathy L. Cripps*

and British Columbia. Limber pine occurs in almost all of these locations plus the Middle Rockies; it is extensive on the Front Range of Colorado.

Subterranean *Rhizopogon* species ("pogies") are excavated by mammals for food. *Photo by Don Bachman*

Mushroom hunting can be challenging in these high, dry habitats, and this sport is certainly not for those who may need the gratification of actually finding mushrooms. But roaming through the gnarled and twisted pines on a blue-sky day can be reward enough and finding mushrooms is the added bonus. The forests usually do not melt out until sometime in July, and then mushrooms can be found around melting snowbanks. In late summer, there is usually a lull in fungal fruiting due to dry conditions, but if the fall monsoons come, mushrooms can be collected into October. These forests host *Suillus* species found only under 5-needle pines, such as *Suillus sibiricus*, *S. subalpinus*, and *S. tomentosus* variety *discolor*. *Boletus edulis* can occur in the forests in Montana, Wyoming, and Idaho. True subterranean fungi such as the false truffle *Rhizopogon* ("pogies") can be located just below cracks in dry soil or as crumbs left in front of squirrel diggings. Many mammals, including squirrels, deer, and bear dig up and eat these fungi—and disperse their spores.

Some mushroom genera that normally fruit above ground avoid dry, cold conditions at these high elevations by remaining underground. We have found fully mature Amanitas, Cortinarii, and Russulas fruiting just below the soil surface in whitebark pine forests.

At the highest elevations near the tree line, whitebark pine becomes stunted into a gnarled krummholz of short stocky trees that form a low dense barrier. This barrier provides habitat and offers protection from the harsh cold, windy environment for rabbits, wild mountain sheep, and small mammals. The leeward side is often a good place to look for mushrooms since it is more likely to hold moisture.

Mature mushrooms fruiting just beneath the soil surface. *Photos by Cathy L. Cripps*

Krummholz whitebark pine on the Hell-roaring Plateau, MT. *Photo by Cathy L. Cripps*

Photo by Cathy L. Cripps

Amanita alpinicola Cripps and Lindgren

DESCRIPTION: CAP 4–8 cm across, broadly convex; white to very pale lemon-yellow; greasy, covered with numerous raised, innate white warts in concentric circles. **GILLS** free; white; close, somewhat broad. **STALK** 6–15 × 1–3 cm, club-shaped; white; sometimes with indistinct ring or ring tissue that can look like torn stalk tissue; at the base, a tidy rimmed cup. **FLESH** white; odor not distinctive. **SPORE PRINT** white.

ECOLOGY: Found in high-elevation whitebark pine forests in Montana and known from a few places in the Pacific Northwest. The fully opened fruiting bodies are often buried in the soil in these high, dry habitats. It is mycorrhizal with whitebark pine and possibly other western 5-needle pines.

OBSERVATIONS: *alpinicola* for its high-elevation habitat and pine-loving nature. Alex Smith gave this species the provisional name "*A. alpina*" but there is already an *Amanita alpina* Contu, which is truly alpine and with willows. *Amanita alpinicola* often fruits underground and can be located by carefully looking for cracked soil in open barren understories. Its toxicity is not known. It is similar to *A. gemmata*, which is likely a complex; here, one species is sorted out.

Photo by Cathy L. Cripps

Hygrophorus gliocyclus Fries

DESCRIPTION: CAP 3–7 cm across, shallow convex or almost flat with a curled-under margin; pale cream; smooth, very slimy. **GILLS** go slightly down the stalk and are well-spaced; thick; pale yellow-peach; they are covered with a slimy veil in young specimens. **STALK** 2–3.5 × 1.5–2.5 cm, tapered toward the base; smooth, very slimy, also from the glutinous veil; pale cream. **FLESH** white; firm; taste mild, odor not detected. **SPORE PRINT** white.

ECOLOGY: Occurring in half-buried clusters in conifer forests, including those of whitebark, lodgepole, and ponderosa pine in addition to those of mixed spruce-fir; typically fruiting in summer or fall; also known from other areas of the western United States. The larger *Hygrophorus* species, like this one, are assumed to be mycorrhizal.

OBSERVATIONS: *gliocyclus* for its gluelike qualities; try sticking the cap to the palm of your hand to confirm the identification! This species is recognized by its squat, robust, pale cream fruiting bodies. It could be confused with *Hygrophorus subalpinus* (Pg 196), which fruits in early spring, but that snowbank species is pure white and not as sticky. Other pale *Hygrophorus* species are less robust. Interestingly, Alex Smith suggests *H. gliocyclus* is edible and choice if the slime is wiped off, but we have no experience with the edibility of this fungus.

Photo by Cathy L. Cripps

Tricholoma moseri Singer

DESCRIPTION: CAP 1–3 cm across; conic-convex with turned-in margin, becoming almost flat; gray-brown, with radial fibrils that darken to black; margin with a cobweb of white fibrils when young. **GILLS** attached, narrow, close to subdistant; whitish, or pale gray. **STALK** 2–5 × 0.3–0.6 cm, equal; white or pale gray; longitudinally fibrous. **FLESH** white or pale gray; odor faint or mealy. **SPORE PRINT** white.

ECOLOGY: Mycorrhizal with many different conifers and especially with young trees but also occurring in mature forests. We see it most often in whitebark and limber pine forests in Montana. It fruits in small groups in late spring and is often one of the first mushrooms to come up after the snowbank fungi depart in early July.

OBSERVATIONS: *moseri* for Meinhard Moser, an Austrian mycologist. Like *Tricholoma myomyces*, it is one of a few small gray Tricholomas that have a slight cobweb of white fibrils on the cap margin when young; *T. moseri* fruits in spring while *T. myomyces* prefers fall. There are many "small, gray" Tricholomas, but the two mentioned above are typical of conifer forests.

Photo by Cathy L. Cripps

Cortinarius pinguis (Zeller) Peintner and M. M. Moser

DESCRIPTION: CAP 2–4 cm across, like a closed-up mushroom that does not open; light yellow-brown, wood-brown, streaked; slimy; with a puckered margin that joins the stalk. **GILLS** (gleba) consisting of a gelatinous convoluted maze; brown, rusty brown; mostly totally enclosed. **STALK** 2–4 cm × 0.5–1.5 cm; half inside the cap and half protruding from cap; smooth, slimy; cream or with yellow-brown tints; slightly swollen at base where there can be a small purple gelatinous cup. **FLESH** solid; cream-colored; odor sweet, oily, or musky-fermented. **SPORES** brown.

ECOLOGY: Buried or almost buried in duff in whitebark pine forests mixed with spruce, or in spruce-fir forests, from July to October. The top of the cap is usually just below the soil surface and is often nibbled by mammals that are attracted by its odor. It is a mycorrhizal fungus of the western United States.

OBSERVATIONS: *pinguis* or "fatty" for its gelatinous slimy covering. It was called *Thaxterogaster pinque* until DNA analysis confirmed it is a *Cortinarius* that does not open (secotioid). Squirrels eat the fungus and the spores are spread in their pellets. However, just because squirrels love it is not a reason to even consider it for eating. Most fungi in the genus *Cortinarius* should not be considered as human food.

Photo by Cathy L. Cripps

Suillus sibiricus (Singer) Singer

DESCRIPTION: CAP 1.4–5 cm across, convex; typically bright yellow with speckles of red and larger red-brown patches toward the margin; smooth, sticky; margin rimmed with white veil remnants when young. **PORES** dull golden yellow; somewhat large, 1–2 mm broad, and radially arranged. **STALK** 2.5–3.5 × 1 cm, mostly equal or narrower at the base; yellow above the ring zone and whiter below, with hints of pink especially toward the base; covered with numerous large black-brown glandular dots. **RING** white, sticky, distinct in young fruiting bodies and later leaving white tissue at the cap margin. **FLESH** dingy white or yellow, can bruise brown; odor not distinctive. **SPORE PRINT** brown.

ECOLOGY: Mycorrhizal with 5-needle pines in the Rocky Mountains, fruiting in late fall from August to October. Common with whitebark and western white pine in the Northern Rockies and reported with limber pine in Montana and Colorado. This species also occurs in stone pine (5-needle) forests in Europe and Asia.

OBSERVATIONS: *sibiricus* for the first description by Rolf Singer who found it associated with Siberian stone pines. It is recognized by the yellow cap with red patches, white tissue at cap margin, and black dots on the stalk. It is similar to the eastern *Suillus americanus*, which occurs with eastern white pines, and they may be the same species. The krummholz form of *S. sibiricus* is much stouter than lower elevation forms. Handling can result in a slight skin rash, so *S. sibiricus* is not recommended for consumption.

Photo by Cathy L. Cripps

Suillus tomentosus var. *discolor* A. H. Smith, Thiers, and O. K. Miller

DESCRIPTION: CAP 3–8 cm across, shallow convex; golden brown, orange-brown; covered with flattened brown fibers or scales pressed to the surface, slightly sticky. **PORES** 1 mm in size, radially arranged, dull gold with slight olive tint. **STALK** 2–3 × 3–4 cm; slightly club-shaped; dingy gold, covered with dark brown-black glandular dots, slightly pink at the base. **FLESH** turning bright or faint blue, especially just above the pores and at the top of the stem when cut open, and pale orange in the stalk; odor faintly fungoid or fruity. **SPORE PRINT** brown.

ECOLOGY: Mycorrhizal with 5-needle pines and reported with whitebark, western white, and limber pines in Idaho, Montana, and Wyoming and with limber pine in Colorado. It fruits from summer to late fall, July through September.

OBSERVATIONS: *tomentosus* for the textured cap and *discolor* for the dark color. This variety is similar to *Suillus tomentosus* variety *tomentosus* (Pg 135), which is common with 2–3-needle pines such as lodgepole. It differs by its darker coloration and scalier cap; both varieties turn blue when cut open, but the oxidation reaction is weaker in variety *discolor* and the stalk flesh is typically pale orange. Some people eat the variety *tomentosus* although the flavor is not prized, but we have no information on the edibility of variety *discolor*.

Photo by Cathy L. Cripps

Calbovista subsculpta Morse ex M. T. Seidl

DESCRIPTION: FRUITING BODY medium-sized, 10 × 15 cm across, round to flattened; white, becoming brown especially between patches with age; surface usually covered with flattened scales but sometimes with pointed gray-tipped warts; surface breaking up into flat patches; skin thick when cut open. **INTERIOR** (gleba) white, becoming brown; large, white, sterile base. **FLESH** firm when young, powdery and brownish at maturity; odor not distinctive. **SPORES** yellowish brown.

ECOLOGY: saprobic; in open grassy places near high-elevation conifer forests, particularly those of spruce-fir and whitebark pine. Fruiting in summer and fall from July to August; common throughout the mountains of western North America.

OBSERVATIONS: *subsculpta* for its sculptured surface, which lacks the tall pointed warts of *Calvatia sculpta*. This baseball-sized puffball has a hefty meaty quality when fresh and is one of the better-tasting puffballs. It is smaller than the giant puffballs and larger than most other puffballs. *Gastropila fumosa* (Pg 220), which is also found at high elevations is smoother and more gray; *Calvatia cretacea* (Pg 236) is smaller with a finer pattern on its surface. Microscopic differences (thornlike branches) inside the gleba distinguish *C. subsculpta* from *Calvatia/Lycoperdon* species, which instead have long tangles of microscopic branching hyphae inside.

Photo by Cathy L. Cripps

Cronartium ribicola J. C. Fischer

DESCRIPTION: FRUITING STRUCTURES scattered as bright orange pustules erupting through the bark on the main trunk or branches of 5-needle pines; consisting of blisters containing an orange spore powder (use a hand lens).

ECOLOGY: White pine blister rust is specific for 5-needle pines and is a pathogen of whitebark pine, limber pine, and western white pine in the Rockies and the Pacific Northwest. The orange pustules are the aecial stage of the rust fungus, and they are present throughout the summer on trunks and branches of infected pines. The aecia burst open to release masses of reddish orange aeciospores, a color that pertains to the name "rust." These spores must infect an alternate host, before the rust can return to the pine.

OBSERVATIONS: *ribicola* for its alternate host *Ribes*, or gooseberry. Rusts have complicated life cycles and produce up to five different kinds of spores. Some, like this rust, alternate between hosts; here, the tradeoff is between *Ribes* (gooseberry) and 5-needle pines. Recently, species of paintbrush (*Castilleja*) and lousewort (*Pedicularis*) have been found to be additional alternate hosts. Huge efforts were expended to eliminate this rust after it was introduced around 1900 and proceeded to reduce western white pine populations; now it threatens the very survival of whitebark pine as a species.

Photo by Cathy L. Cripps

Gastropila fumosa (Zeller) P. Ponce de León

DESCRIPTION: FRUITING BODY round to ovoid, 4–8 cm across; gray-brown or smoky gray, covered with fine cracks and fissures that reveal the white color beneath; dry; with very thick skin (cut open); short attachment cord at the base. **INTERIOR** flesh pure white, very solid when young, becoming olive-brown and powdery with age; odor unpleasant when mature. **SPORES** brown.

ECOLOGY: Fruiting just above the soil or slightly buried in needle litter in high-elevation conifer forests of western North America. It is particularly common in spruce-fir and whitebark pine forests, fruiting in summer and fall from late July into September.

OBSERVATIONS: *fumosa* for its foul odor. Also known as *Calvatia fumosa,* this fungus is recognized by its hard flesh, thick skin, and foul odor when mature. *Scleroderma* species share the first two characteristics but are some shade of purple-black inside even when young; they are toxic. When half-buried, *G. fumosa* can be confused with Rhizopogons, which do not have the thick rind. Unlike other small to medium puffballs, *G. fumosa* lacks a pore for spore dispersal. *Gastropila fumosa* should not be considered for the table, thus making the answer to the question "Are all puffballs edible?" a resounding: No!

Photo by Cathy L. Cripps

Rhizopogon evadens A. H. Smith

DESCRIPTION: FRUITING BODY 2.5–5 cm across; mostly buried in the ground; round and lumpy like a small tuber; dingy white with pinkish tints, staining more pink on handling; surface a bit rough, covered with fine pink rhizomorphs (cords). **INTERIOR** (gleba) solid when young; pure white, turning grayish olive when mature; consisting of a fine labyrinth of pores (use a hand lens); odor not distinctive or faintly musky. **SPORES** brown, not turning blue in Melzer's iodine solution.

ECOLOGY: Occurs belowground in whitebark pine, western white pine, limber pine, and possibly bristlecone pine forests at high elevations; also with lodgepole and ponderosa pine at lower elevations; fruiting in late fall. Squirrels, bears, and deer locate the fruiting bodies by their odor for a tasty meal. Spores are spread in animal droppings; they germinate and reassociate with tree roots as a mycorrhizal fungus.

OBSERVATIONS: This species is very common in whitebark and limber pine forests and is recognized by its pink tints and rhizomorphs. *Rhizopogon milleri* is also common with 5-needle pines, but it has amyloid spores and the outer skin lacks pink tints. *R. rubescens* stains reddish in the outer surface when injured; it is mycorrhizal with whitebark as well as lodgepole pine.

ALPINE

Photo by Andy Hogg

THE ALPINE

The true alpine is the open windswept land above the tree line on high mountain tops. Above the subalpine forests, which end in a krummholz of deformed trees, lies an exposed mosaic of low vegetation types—the alpine tundra. This is the coldest of all Rocky Mountain habitats where snow covers the landscape and lakes remain frozen most of the year. In late June, vegetation begins to reappear as the temperatures rise and the snow melts, but after the short growing season in July and August, the landscape quickly gives way to winter again. Snow and freezing temperatures can occur any time of year. Despite these limitations, the Rocky Mountain alpine is one of the most spectacular habitats in terms of breathless vistas, rareified air, and the unique diversity of life adapted to these high elevations. In Montana, the alpine zone begins around 10,500 feet, but as the tree line rises to the south, it reaches 12,000 feet in Colorado.

Mountain goats, bighorn sheep, pikas, and marmots are characteristic mammals that make the alpine their home. Mountain goats (*Oreamnos americanus*) standing like

Mountain goats have a white shaggy coat that matches the snow. *Photo by Lee Gillman*

A hoary marmot (*Marmota caligata*) with silvery fur and black nose band. *Photo by Cathy L. Cripps*

white sentinals on rocky crags are a surprise when one catches sight of them. They have a broad range in western North America, although some populations are introduced. Fat yellow-bellied (*Marmota flaviventris*) and hoary marmots (*M. caligata*) waddle across the alpine landscape and can be heard giving high-pitched warning whistles before escaping to their rocky underground homes. Cute round-eared pikas (*Ochotona princeps*) sound warning "eeks" that echo through the thin alpine air. These cousins of rabbits are true cold lovers, and their metabolism prevents them from surviving in warmer low-elevation habitats. Pikas are industrious creatures; they continually gather vegetation in summer, which they stuff into rock piles for winter food.

American pika has small round ears. *Photo by Andy Hogg*

A white-tailed ptarmigan blending in with lichen-covered rocks. *Photo by Lee Gillman*

Low cushion plants such as alpine forget-me-not and moss campion are common in the alpine. *Photos by Cathy L. Cripps*

Birds pass through the alpine during their migrations, an occasional hawk or raven might careen high above, but it is the white-tailed ptarmigan (*Lagopus leurura*) that inhabits these cold climes year around. These cold-loving birds turn white in winter, blending with the snowscape, and brown in summer, matching the mottled lichen-covered rocks.

The alpine landscape is a patchwork of open grasslands, wet and dry meadows, *Geum* turfs, and shrublands. In summer, the alpine comes alive with the flowering of low-cushion plants such as alpine forget-me-nots (*Eritrichium nanum*), pale phlox (*Phlox pulvinata*), and pink moss campion (*Silene acaulis*). Grasslands and meadows abound with paintbrush, bistorts, white arctic gentians, skypilot, yellow alpine avens, and cottongrass.

The alpine is characterized as treeless, but miniature forests of dwarf willows blanket the alpine soil in wet areas. Two of the most common, *Salix arctica* and

Dwarf willows *Salix arctica* and *S. reticulata* are only a few inches in height. *Photos by Cathy L. Cripps*

Alpine morel (*Morchella* sp.) nestled in alpine vegetation and a basidiolichen *Arrhenia* in moss. *Photos by Cathy L. Cripps*

S. reticulata, are less than a few inches high. Mountain dryad (*Dryas octopetala*) is another mat plant that, together with willows, comprise the main ectomycorrhizal plants in the Rocky Mountain alpine. Mushrooms associated with these mat plants can tower over their plant hosts reversing the typical situation below the tree line where trees rise above their fungal mycorrhizal allies.

The alpine hardly seems like a place to look for mushrooms, but they are there, nestled beneath willows, tucked into mosses, and hidden in the grasses. Baskets of edibles are rare unless a sprout of large puffballs or meadow mushrooms appear, but the search for tiny fungal gems is worthwhile because alpine mushrooms are truly special. Basidiolichens are common in the alpine and are a unification of green algae and mushroom-producing fungi. The algae use sunlight, air, and water to produce food for both organisms in this symbiosis. The closest relatives of the alpine mushrooms in the Rocky Mountains lie far to the north in arctic areas of Alaska, Canada, Greenland, Iceland, Eurasia, and in alpine areas of Europe and Asia.

Melanoleuca cognata in an alpine meadow on Niwot Ridge, Colorado. *Photo by Cathy L. Cripps*

Arctic-alpine fungi exist only on the very tips of the Rocky Mountains along the backbone of North America up to and around the Arctic Circle. In Montana and Wyoming, alpine mushrooms can be found on the Beartooth and surrounding plateaus. Farther south in Wyoming, look for them in the Wind Rivers and the Medicine Bow Ranges. In Colorado, they occur on high passes such as Loveland and Independence and in the high cirques of the Front Range, Sawatch Range, and the San Juan Mountains to name a few habitats above the tree line.

Photo by Cathy L. Cripps

Amanita groenlandica forme alpina Cripps and Horak

DESCRIPTION: CAP 4–8(12) cm across, broadly convex, smooth; pale salmon-buff; "creamsicle" color, white when young, silvery metallic brown with age; with large patches of cream-colored velar tissue scattered in the center; margin slightly pleated. **GILLS** free, crowded; white or tinted pale orange; edges fringed. **STALK** 4–10 cm long × 1–3 cm wide, gradually enlarging toward base, with an "adder-pattern" on the surface; white with pale orange or gray tints; without a ring; base enclosed in a flaring, whitish loose sac tinted pale orange or gray. **FLESH** white; odor fruity or unpleasant with age. **SPORE PRINT** white.

ECOLOGY: Occurring in alpine habitats above the tree line with its mycorrhizal partners *Salix reticulata*, *S. arctica*, and the shrub willow *S. glauca*. Large fruitings are common on the Beartooth Plateau in Wyoming and Montana in August, but it is not yet recorded from Colorado. It is also known from Greenland.

OBSERVATIONS: *groenlandica* for the first description from arctic Greenland and the alpine form was later recognized in Montana. There are few Amanitas in the alpine zone with willows. This one is recognized by its "creamsicle" color, lack of a ring, and robust stature. The more famous arctic *Amanita nivalis* is smaller, more delicate, light gray, and occurs in the alpine cirques of the San Juan Mountains, Sawatch Range, Loveland Pass, and Rollins Pass in Colorado.

Photo by Vera S. Evenson

Arrhenia auriscalpium (Fries) Fries

DESCRIPTION: CAP very tiny, 3–6(15) mm across; erect, spoon-shaped, thin-fleshed, pale brown to gray-brown; somewhat translucent, often with a lobed margin that is curled over. **GILLS** reduced to a few shallow, widely spaced ridges that branch toward the margin; same color as cap or a bit lighter. **STALK** 2–4 mm long × 1–2 mm wide, attached to one side of cap; smooth or minutely velvety; pale brown to gray-brown. **SPORE PRINT** white.

ECOLOGY: Found in true alpine habitats at high elevations and also known from the arctic near sea level. Occurring on open organic sun-exposed soil near minute mosses or the white worm lichen (*Thamnolia subuliformis*). Rare or rarely noticed.

OBSERVATIONS: *auriscalpium* for the ear shape of this tiny fungus. The similar and more common *Arrhenia lobata* occurs in alpine as well as subalpine habitats; it is usually attached to mosses—more rarely, twigs—and while it also has reduced ridges instead of gills, it is larger, more lobed, and lacks a distinct stalk. *Arrhenia auriscalpium* is considered an indicator of true arctic and alpine habitats and is reported from Greenland, Iceland, the Alps, Scandinavia, Svalbard, Russia, and Alaska. It is also recorded from Loveland and Independence Passes in Colorado and from Alaska in North America.

Photo by Ed Barge

Omphalina rivulicola (J. Favre) Lamoure

DESCRIPTION: CAP 1–3 cm across, deeply funnel-shaped; liver-brown when fresh, drying pale brown to almost white; smooth, greasy; margin rolled under at first, wavy with age and slightly pleated. **GILLS** well-separated, running down the stalk; pale cream with edges that darken. **STALK** 2–5 cm × 1–4 mm, long and thin, smooth; pale brown, often white at the base. **FLESH** watery cream; thin in cap; odor absent. **SPORE PRINT** white.

ECOLOGY: Occurring in alpine habitats, scattered in moss in wet areas near streams and lakes. This delicate funnel-shaped mushroom can be common along streams, but in years of high and fast runoff when streambanks are scoured we found it almost absent. It occurs at least from Colorado to Montana and is known from other arctic-alpine habitats, including Iceland, Svalbard, Greenland, and the Alps.

OBSERVATIONS: *rivulicola* for its river-loving habit. The funnel shape is reminiscent of a *Clitocybe*, but it is more delicate and pales in color on drying. There are many other funnel-shaped (omphalinoid) fungi in arctic-alpine habitats (*Omphalina*, *Arrhenia*) but most are much smaller. While some occur in mosses, many are found on open mudflats where they make their living as Basidiolichens (mushroom-producing lichens).

Photo by Cathy L. Cripps

Rhizomarasmius epidryas (Kühner ex A. Ronikier) A. Ronikier and M. Ronikier

DESCRIPTION: CAP 5–10 mm across, convex with a slightly depressed center; variable in color, from red-brown to pale yellow-brown, with dark red-brown center; margin pleated. **GILLS** broadly attached or running slightly down the stalk, a bit thick; pale cream; sparse (only 15 to 20 gills). **STALK** 5–20 mm long × 1–2 mm wide; equal, a bit curved; dark red-brown to black; minutely velvety over the whole length (use a hand lens). **FLESH** thin, but tough and pliable; odor indistinct. **SPORE PRINT** white.

ECOLOGY: Apparently saprobic on dead parts of *Dryas* (small mat plants with scalloped leaves as shown on Pg 227) in arctic and alpine habitats. It is reported from a few alpine locations in the Rocky Mountains, such as Loveland Pass; also known in Canada, Alaska, Greenland, Scandinavia, Romania, Russia, and the Alps.

OBSERVATIONS: *epidryas* for its location on dead parts of *Dryas,* a plant in the rose family. This strict association and its tough black velvety stalks make *R. epidryas* easy to recognize. Once called *Marasmius epidryas* and briefly *Mycetinis,* molecular analysis placed it in the new genus *Rhizomarasmius*. It does not appear to be parasitic even though it is attached to *Dryas*.

Photo by Cathy L. Cripps

Russula nana Killerman

DESCRIPTION: CAP 2–5 cm across, convex to almost flat, smooth, a bit greasy; bright cherry-red, paling to white particularly in the center; a few grooves at the margin or not. **GILLS** attached; white, graying slightly with age. **STALK** 2–3.5 × 1–1.5 cm, gradually larger toward base or not, smooth, fibrous, white. **FLESH** white but graying a bit, brittle especially in stalk; odor absent, but taste can be radish-hot in the morning when fresh, otherwise mild depending on the weather. **SPORE PRINT** white.

ECOLOGY: One of the most common species in the Rocky Mountain alpine and its bright red cap stands out against the alpine vegetation. It is mycorrhizal with dwarf willows *Salix reticulata* and *S. arctica*. A well-known component of arctic-alpine habitats, *R. nana* has a circumpolar distribution; one of the first fungi to fruit in the alpine (as early as July); it is recorded from alpine sites all over Montana, Colorado, and Wyoming.

OBSERVATIONS: *nanus* for a dwarf or nano-form and this species looks like a miniature *Russula emetica*. As it ages, the cap turns white from the center until only the margin remains red. There are other reddish Russulas in the Rocky Mountain alpine. *Russula laccata* (formerly *R. norvegica*) has a dark maroon cap with an almost black center; it has a hot taste. The larger orange-red *R. pascua* smells fishy and is like a miniature shrimp *Russula* (*R. xerampelina*), but the edibility of the alpine form is untested. *Russula nana* was first reported from the Rocky Mountains in 1987 when well-known Austrian mycologist Meinhard Moser took a trip over Beartooth Pass.

Photo by Cathy L. Cripps

Cortinarius favrei D. M. Henderson

DESCRIPTION: CAP 1–4 cm across, conic-convex or almost flat with a turned-down margin; sticky, glutinous; caramel color (orange-brown), sometimes darker in the center and more yellow at the margin, which is covered by a glutinous veil. **GILLS** attached; pale cream at first, foxy brown when mature, occasionally with a lavender tint. **STALK** 2–6 by 0.5–1 cm, equal or tapered to a point; white; covered with glutinous material; smooth above the glutinous ring where rust-colored spores stick. **FLESH** white; often hollow in stalk; odor absent. **SPORES PRINT** rusty brown, although gills can be pale.

ECOLOGY: Common in the alpine zone and one of the first alpine fungi to appear in late July. Mycorrhizal and found scattered near dwarf willows in Colorado, Wyoming, and Montana. This famous alpine fungus is well-known from many arctic and alpine habitats around the northern hemisphere.

OBSERVATIONS: *favrei* for the father of alpine mycology Jules Favre, who studied the alpine fungi of the Swiss Alps. Also called *C. alpinus*. This *Cortinarius* is recognized by its conspicuous bright orange cap and slippery fruiting bodies. Our similar but larger local fungus, *C. absarokensis*, was named for the mountains near the Beartooth Plateau in Montana by Meinhard Moser. *C. absarokensis*, a primarily North American species, is larger (up to 12 cm across), the stalk is tough, woody, and black at the base, and it occurs with shrub (not dwarf) willows.

Photo by Cathy L. Cripps

Leccinum rotundifoliae (Singer) A. H. Smith, Thiers, and Watling

DESCRIPTION: CAP 6–9 cm across, broadly convex, slightly greasy or dry; cream, pale brown, or pale yellow-brown; smooth or becoming finely cracked in a pattern. **TUBES** indented at stipe; pores tiny (2/mm); white to pale gray, staining brownish. **STALK** 7–13 × 2–3 cm, enlarged toward base; whitish cream, with pinkish tints; covered with pale grayish brown rough scales. **FLESH** white, staining slightly pinkish or wine color, but not blue; odor slightly sweet. **SPORE PRINT** yellow-brown.

ECOLOGY: One of the few boletes that occur in the alpine or upper subalpine with dwarf birch, especially *Betula glandulosa/nana*, and apparently restricted to birch. It is known from Greenland, Iceland, and Alaska. Birch is rare in the Rocky Mountain alpine. This species occurs with the few birch shrubs found on the Beartooth Plateau in Montana, and there is one report from Idaho.

OBSERVATIONS: *rotundifoliae* for its association with *Betula rotundifolia*, round-leafed dwarf birch. Singer first described this species from the Altai Mountains in Asia. Recognized by its pale cap and habitat with dwarf birch, this species is considered close to the orange-capped *L. scabrum* also known from arctic-alpine habitats with birch. Molecular evidence suggests *L. rotundifoliae* is a separate species; the mushroom shown in the photograph is a genetic match to *L. rotundifoliae* in Europe. It could be confused with the white-capped *L. holopus*, which is not an alpine species.

Photo by Cathy L. Cripps

Calvatia cretacea (Berkeley) Lloyd

DESCRIPTION: FRUITING BODY 4–8 cm across by 3–6 cm high; round, flattened ovate, or sometimes pear-shaped; thin-skinned; dingy white to cream, dark brown with age; surface pattern of very fine spines in clusters; spines eroding and surface developing a fine-patterned cracked appearance. **INTERIOR** flesh white, then yellow-brown or dark brown when powdery with spores; sterile base present but can be minimal; odor somewhat unpleasant. **SPORES** yellowish brown.

ECOLOGY: Scattered in open grasslands, in moss, or in the open at high elevations where it is a decomposer of recalcitrant organic material in the soil. One of the most common puffballs in arctic-alpine habitats; it has been reported from Alaska, Montana, Colorado, and Canada in North America and has a circumpolar distribution.

OBSERVATIONS: *cretacea* for its chalky appearance. Also called *Lycoperdon cretaceum*, it is a medium-sized puffball with very fine spines that cluster together in young specimens. It is in the group of puffballs that lack a pore and simply crack open at maturity for spore dispersal. Its thin skin helps distinguish it from other high-elevation puffballs such as *Calbovista subsculpta*, which has truncate warts (Pg 218), *Gastropila fumosa*, which is gray and smoother (Pg 220), and *C. (Lycoperdon) subcretacea*, which has gray warts; all three have thick skins. The giant western puffball *C. booniana* (Pg 41) can sometimes reach the low alpine, but is much larger. Of these, only *C. booniana* and *Calbovista subsculpta* are considered good edibles.

MACROFUNGI GROUPED BY MORPHOLOGY

Species are grouped by morphology so you can quickly determine whether a particular fungus or mushroom is included in the book; this list also can be used as a key to the main groups.

Light-spored Gilled Mushrooms

Amanita
Photo by Cathy L. Cripps

1. Gills clearly free from the stalk

Amanita species: cup at base of stalk [pp. 21, 53, 75, 105, 165, 166, 212, 229]

Lepiota clypeolaria: without cup at base of stalk [p. 170]

1. Gills attached or running down the stalk, or stalk absent **2**

Pleurotus
Photo by Cathy L. Cripps

2. Stalk attached to one side of cap or absent

Arrhenia auriscalpium: very tiny (mm); spoon-shaped; on alpine soil [p. 230]

Lentinellus montanus: stalk absent; gills saw-toothed; on wood [p. 197]

Neolentinus ponderosus: big; stalk off center; gills saw-toothed; on wood [p. 79]

Pleurotus species: stalk short or absent; gills go down stalk; on wood [pp. 57, 110]

2. Stalk attached to center of cap **3**

Flammulina
Photo by Cathy L. Cripps

3. Mushrooms with thin stalks; on wood, needles, twigs, cones, or *Dryas*

Flammulina populicola: black velvety stalks; in clusters on wood; on aspen [p. 106]

Gymnopus perforans: tiny, black horsehair stalk; on conifer needles [p. 76]

Heliocybe sulcata: pleated cap; saw-toothed gills; on aspen wood [p. 108]

Mycena acicula: bell-shaped, reddish; at base of wood, in clusters; riparian [p. 56]

Mycena overholtsii: bell-shaped, grayish; in clusters on wood; snowbanks [p. 198]

Rhizomarasmius epidryas: on *Dryas* in the alpine [p. 232]

Strobilurus albipilatus: on conifer cones [p. 97]

Tubaria furfuracea: on aspen debris such as dead leaves and twigs [p. 112]

Xeromphalina species: small; gills go down stalk; on wood, needles, or debris [pp. 81, 175]

3. Mushrooms with thicker stalks; on the ground **4**

Flocculaira
Photo by Cathy L. Cripps

4. Stalk with a ring or ring zone

Armillaria solidipes: yellow-brown cap; in clusters at base of trees [p. 8]

Cercopemyces: whitish granular cap; under mountain mahogany [p. 38]

Cystodermella cinnabarina: orange or brown, granular cap [p. 127]

Floccularia species: gray, yellow, white cap; stalk scaly [pp. 93, 94, 107]

Hygrophorus olivaceoalbus: slimy; grayish; with waxy gills [p. 167]

Hygrophorus subalpinus: white; meaty; near snowbanks [p. 196]

Tricholoma species: a few species have rings [pp. 131, 132]

4. Stalk without a ring **5**

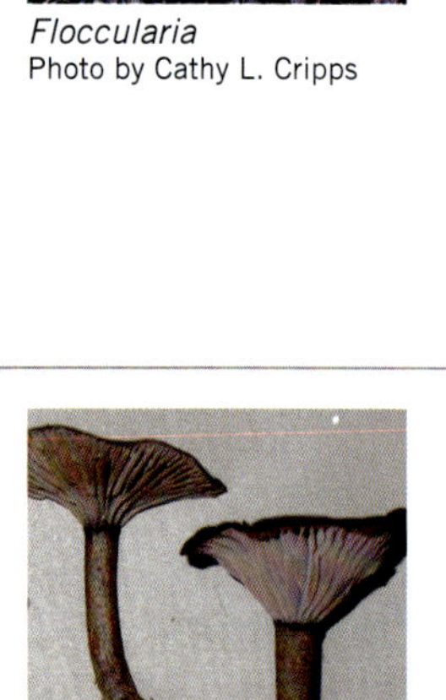

Neohygrophorus
Photo by Cathy L. Cripps

5. Gills go down the stalk; funnel-shaped

Hygrophorus species: thick-fleshed; gills wide apart, waxy [pp. 77, 168, 195, 213]

Lactarius species: gills exude "milk" when cut, or flesh turns green [pp. 55, 95, 109, 129, 169]

Leucopaxillus albissimus: white, chalky; long stalk; tough [p. 78]

Neohygrophorus angelesianus: small; violet-brown; near melting snowbanks [p. 199]

Omphalina species: small, thin-fleshed, fragile; often translucent [p. 231]

Russula brevipes: big; white; dense flesh; short stalk, half buried [p. 130]

Clitocybe species: white, tan, orange, or bluish caps; with pungent odors

Russula
Photo by Cathy L. Cripps

5. Gills attached, not going down the stalk; not funnel-shaped

Clitocybe species: some have gills that do not go down the stalk; near melting snowbanks [pp. 193, 194] or in open sagebrush [p. 22]

Laccaria species: orange, with thick pink gills [pp. 54, 128]

Marasmius oreades: small; stalks thin, tough, or brittle; in grass [p. 23]

Mycena pura: lavender to whitish with white spores [p. 171]

Russula species: red, green, purple, etc.; flesh fragile, crumbly, stalk chalklike [pp. 80, 96, 111, 130, 172, 233]

Tephrocybe atrata: small; gray-brown; on burned ground [p. 147]

Tricholoma species: small to big, fleshy; white, gray, yellow, brown caps [pp. 58, 59, 173, 174, 214]

Pink-spored Gilled Mushrooms

Entoloma
Photo by Michael Kuo

1. Gills attached to stalk; on wood or on the ground

Entoloma species: often with whitish or pinkish gills; on ground [p. 113]

Volvariella
Photo by Cathy L. Cripps

1. Gills free from stalk; often pinkish; on wood or on the ground

Pluteus: without a cup at the base of the stalk

Volvariella: with a small cup at the base of the stalk

Dark-spored Gilled Mushrooms

Coprinus
Photo by Ed Barge

1. With jet-black spores; gills turning to ink in some (inky caps)

Coprinopsis variegata: felty; white; egg- to bell-shaped cap [p. 60]

Coprinus comatus: shaggy; white; egg- to bell-shaped cap [p. 61]

1. Spores not jet black 2

Pholiota
Photo by Cathy L. Cripps

2. On wood or in clusters at the base of trees, or on burns

Pachylepyrium carbonicola: small; rusty brown spores; on burns [p. 148]

Pholiota highlandensis: smooth; orange cap; scattered on burns [p. 149]

Pholiota squarrosa: scaly; in dense clusters at base of trees [p. 117]

Psathyrella species: bell-shaped; white brittle stalks; in clusters; on burns or not [pp. 62, 118, 150]

Stropharia riparia: cream cap with purple-black spores; on wood [p. 63]

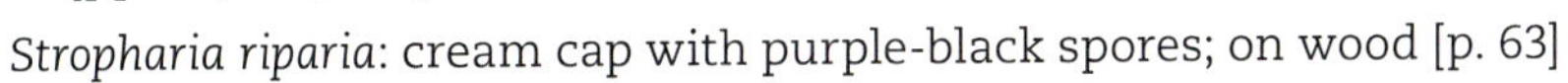

2. On the ground (one on dung). 3

Agaricus
Photo by Cathy L. Cripps

3. Gills free (pink, brown, or gray); spores chocolate brown

Agaricus species: caps often whitish; ring or ring zone; often in grass [pp. 24, 25, 176]

3. Gills attached or running down the stalk. 4

Gomphidius
Photo by Cathy L. Cripps

4. Gills running down the stalk; funnel-shaped

Gomphidius species: greasy; gills white, gray, black; spores smoky blackish [p. 98]

Paxillus species: dry cap; gills eventually brownish with brownish spores [p. 116]

Cortinarius
Photo by Cathy L. Cripps

4. Gills attached but not running down the stalk; not funnel-shaped

Cortinarius species: rusty brown spores; cobwebby veil [pp. 114, 177, 178, 200, 215, 234]

Deconica coprophila: bell-shaped, thin stalk; purple-black spores; on dung [p. 19]

Hebeloma species: brown spores; some have odor of radish or raw potato [p. 115]

Inocybe species: small, pointed caps; brown spores; odor often spermatic

Stropharia: purple-black spores; ring or ring zone [p. 133]

Suillus
Photo by Cathy L. Cripps

Boletes: fleshy cap and stalk; with pores/tubes

Boletus species: large cap; stalk bulbous, smooth or with tiny veins [pp. 82, 164, 176, 179]

Leccinum species: large cap; large rough stalk; flesh usually graying when cut [pp. 119, 235]

Suillus species: small slimy cap; small stalk, some with dots [pp. 83, 99, 134, 135, 216, 217]

Cantharellus
Photo by Ed Barge

Non-gilled Fungi

1. Club or coral-shaped, or stalk with ridges or wrinkles

Alloclavaria purpurea: purple; pointed finger-shaped; in dense groups [p. 181]

Artomyces pyxidatus: whitish branched coral; on aspen or cottonwood [p. 120]

Cantharellus species: apricot color, funnel-shaped; with ridges [p. 137]

Clavariadelphus species: orange club; sometimes wrinkled [p. 182]

Polyozellus multiplex: purple; funnel-shaped; in compound clusters [p. 184]

Also see *Spathularia flavida* under Ascomycota

Clavariadelphus
Photo by Ed Barge

2. **Fleshy fungi with spines or teeth**

   Auriscalpium vulgare: small cap; brown teeth and stalk; on conifer debris [p. 84]

   Hericium coralloides: white branched spines; no stalk; on wood [p. 64]

   Hydnum repandum: small; pale orange caps; teeth; small stem [p. 183]

   Sarcodon imbricatus: large brownish scaly caps; teeth; fleshy stem [p. 185]

Hydnum
Photo by Ed Barge

3. **Flesh tough; with pores (one with false gills); most shelflike on wood (polypores)**

   Albatrellus confluens: pores very tiny and confluent with stalk; on ground [p. 180]

   Cryptoporus volvatus: woody knob, pores covered; on conifers [p. 100]

   Fomitopsis species: woody, shelflike; with red belt or rosy color [pp. 85, 138]

   Gloeophyllum sepiarium: woody, shelflike; with false gills; on conifers [p. 139]

   Phellinus tremulae: woody, shelflike; with golden brown flesh; on aspen [p. 103]

   Polyporus cryptopus: cap, pores, and stalk; on woody roots in prairie [p. 26]

   Porodaedalea pini: woody, shelflike; with golden flesh; on conifers [p. 86]

   Pycnoporellus alboluteus: soft shelflike; orange; on wood near snowbanks [p. 202]

   Pycnoporus cinnabarinus: orange; shelflike; orange pores [p. 46]

   Trichaptum biforme: woody, shelflike; with lavender pores; on aspen [p. 121]

   Tyromyces leucospongia: soft shelflike; white; on wood near snowbanks [p. 203]

Fomitopsis
Photo by Cathy L. Cripps

4. **Other, miscellaneous fungi**

   Cronartium ribicola: orange blister rust fungus on pine bark and branches [p. 219]

   Guepiniopsis alpina: yellow jelly fungi on wood (lemon drops) [p. 201]

Guepiniopsis
Photo by Cathy L. Cripps

Puffballs, Their Relatives, and Look-alikes with Roundish Fruiting Bodies

1. **Fruiting body as a closed-up mushroom with a stalk**

   Cortinarius pinguis: brown, slimy cap; stalk buried; conifers [p. 215]

   Montagnea arenaria: white dry cap; black plates; arid habitats [p. 37]

   Podaxis pistillaris: whitish cap; long woody stalk; in dry places [p. 39]

1. **Roundish; not becoming powdery at maturity; buried in ground**

   Rhizopogon species: roundish; with tiny pores inside [pp. 136, 221]

1. **Round; becoming powdery at maturity (one with a stem); not buried**

Podaxis
Photo by O.K. Miller, Jr.

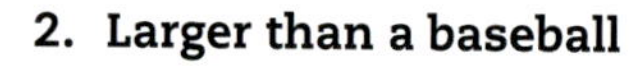

2. Larger than a baseball

Calvatia booniana, C. bovista, C. cyathiformis, Mycenastrum corium [pp. 27, 28, 41, 43]

Calbovista
Photo by Cathy L. Cripps

2. Smaller than a baseball

Bovista pila: outer layer wearing away, inner layer papery [p. 40]

Calbovista subsculpta: thick-skinned; patterned, white [p. 218]

Calvatia cretacea: thin-skinned; high elevations [p. 236]

Disciseda bovista: roundish to flattened, with pore; soil attached to lower part [p. 29]

Gastropila fumosa: thick-skinned; smooth, gray [p. 220]

Geastrum/Myriostoma: outer layer splitting into starlike rays [pp. 42, 44]

Tulostoma fimbriatum: small round sac, pore; woody stem; often buried in sand [p. 45]

Cup Fungi, Earth Tongues, Morels, and False Morels (Ascomycota)

Morchella
Photo by Cathy L. Cripps

1. Fruiting body not cuplike; cap with wrinkles, folds or pits; with a stalk

Gyromitra species: cap with wrinkles, walnutlike; reddish [pp. 140, 205]

Morchella species: caps with pits and ridges; edge attached to stalk [pp. 65, 152, 153]

Verpa bohemica: cap with longitudinal wrinkles; edge hangs free [p. 66]

1. Fruiting body cuplike or cap flattened **2**

Geopyxis
Photo by Cathy L. Cripps

2. With a stalk

Geopyxis carbonaria: orange cup with a stalk; on burns [p. 151]

Helvella acetabulum: brown cup with veined stalk [p. 141]

Plectania nannfeldtii: black cup with a black hairy stalk; near snowbanks [p. 206]

Spathularia flavida: earth tongue; flattened yellow head [p. 87]

2. Without a stalk

Anthracobia melaloma: small; orange; on burns [p. 154]

Caloscypha fulgens: orange-red cup, stains blue [p. 204]

Peziza sublilacina: flattish lavender cup; on burns [p. 155]

Plicaria endocarpoides: brownish olive irregular cup, on burns [p. 156]

Sarcosphaera coronaria: fairly large round lavender cup; under conifers [p. 186]

Scutellinia scutellata: small, red cup with "eyelashes" [p. 187]

ON EATING WILD MUSHROOMS

Observation of the endless diversity, unique ecologies, and astonishing beauty of mushrooms is sufficient sustenance for many mushroom enthusiasts. However, we realize there is also interest in the edibility of wild mushrooms, and information for several well-known choice edibles is provided in this book. We do caution those interested in eating wild mushrooms to keep in mind that *the only way to assess the edibility of a mushroom is to accurately identify it to species*. While this book describes several species for each habitat, it is certainly not comprehensive and "sight identification" is only the first step in the process. In most cases, additional resources will be needed to confirm identifications and to rule out poisonous species. This is particularly true as it pertains to microscopic characteristics, which we have not included, and look-alike species, many of which we do not mention.

Every year serious mushroom poisonings are reported in the Rocky Mountains. If symptoms do occur after eating wild mushrooms, the Rocky Mountain Poison and Drug Center (1-800-222-1222) is a good resource for you and your doctor. We have also included references at the end of this book that contain information on the various groups of poisonous mushrooms.

Basic Cautions for Edible Mushrooms

1. **Do not eat wild mushrooms raw; always cook them first.**
   Some edible species can be toxic and many are not easily digested when eaten raw.
2. **Do not eat old or decaying mushrooms.**
   In the excitement of finding an edible species, this caution is often forgotten.
3. **Eat only a small amount when you are trying a mushroom for the first time.**
   Some people have idiosyncratic reactions to edible mushrooms, and especially to black morels.
4. **Do not overindulge.**
   Fungi are basically made of chitin, a rather indigestible substance for humans.
5. **Caution is advised when drinking alcohol with certain mushrooms.**
   The combination can cause a reaction in certain individuals.
6. **Avoid eating mushrooms from soil contaminated with heavy metals, pesticides, herbicides, and animal waste.**
   Mushrooms can act like little sponges, soaking up residues in the soil.

Before consuming any wild mushroom, make sure it is accurately identified; mycological societies that can help are listed on the North American Mycological Association website: www.namyco.org. The authors and the publisher disclaim any liability resulting from the use of this book and the use of information contained in this book.

FURTHER READING AND REFERENCES BY HABITAT

General Field Guides for the Rocky Mountain Region

Bessette, A. E., D. B. Harris, and A. R. Bessette. 2009. *Milk Mushrooms of North America*. Syracuse University Press, Syracuse, N.Y.

Bessette, A. E., W. C. Roody, and A. R. Bessette. 2000. *North American Boletes*. Syracuse University Press, Syracuse, N.Y.

Beug, M., A. E. Bessette, and A. R. Bessette. 2014. *Ascomycete Fungi of North America*. University of Texas Press, Austin.

Evenson, V. S. 1997. *Mushrooms of Colorado and the Southern Rocky Mountains*. Westcliffe Publishers, Engelwood, Colo.

———. 2015. *Mushrooms of the Rocky Mountain Region: Colorado, New Mexico, Utah, Wyoming*. Timber Press, Portland, Ore.

Gilbertson, R. L., and L. Ryvarden. 1986. *North American Polypores Vol. 1*. Fungiflora, Oslo, Norway.

———. 1987. *North American Polypores Vol. 2*. Fungiflora, Oslo, Norway.

Hesler, L. R., and A. H. Smith. 1963. *North American Species of Hygrophorus*. University of Tennessee Press, Knoxville.

———. 1979. *North American Species of Lactarius*. University of Michigan Press, Ann Arbor.

Laursen, G. A., and R. D. Seppelt. 2009. *Common Interior Alaska Cryptograms*. University of Alaska Press, Fairbanks.

McKnight, K. H., and V. B. McKnight. 1987. *A Field Guide to Mushrooms of North America*. Houghton Mifflin, Boston.

Miller, O. K., Jr., and H. H. Miller. 2006. *North American Mushrooms*. Falcon Guide, Pequot Press, Helena, Mont.

Schalkwijk-Barendsen, H. M. E. 1991. *Mushrooms of Western Canada*. Lone Pine Press, Edmonton, Alberta, Canada.

Smith, A. H. 1972. *The North American Species of Psathyrella*. Memoirs of the New York Botanic Gardens 24: 1–633.

———. 1975. *A Field Guide to Western Mushrooms*. University of Michigan Press, Ann Arbor.

Smith, A. H., V. S. Evenson, and D. H. Mitchel. 1983. *The Veiled Species of Hebeloma in the Western United States*. University of Michigan Press, Ann Arbor.

Smith, A. H., and L. R. Hesler. 1968. *The North American Species of Pholiota*. Hafner Pub. Co., New York.

Smith, A. H., and H. V. Smith. 1973. *How to Know the Non-Gilled Fleshy Fungi*. Wm. C. Brown Inc., Dubuque, Iowa.

Smith, A. H., H. V. Smith, and N. S. Weber. 1979. *How to Know the Gilled Mushrooms*. Wm. C. Brown Inc., Dubuque, Iowa.

Smith, A. H., and H. D. Thiers. 1969. *A Contribution toward a Monograph of North American Species of Suillus*. University of Michigan Press, Ann Arbor.

States, J. 1990. *Mushrooms and Truffles of the Southwest*. University of Arizona Press, Tucson.

Trudell, S., and J. Ammirati. 2009. *Mushrooms of the Pacific Northwest*. Timber Press, Portland, Ore.

Tylutki, E. E. 1979. *Mushrooms of Idaho and the Pacific Northwest: Discomycetes*. University of Idaho Press, Moscow.

———. 1987. *Mushrooms of Idaho and the Pacific Northwest. Vol. 2: Non-gilled Hymenomycetes*. University of Idaho Press, Moscow.

Wells, M. H., and D. H. Mitchel. 1966. *Mushrooms of Colorado and Adjacent Areas*. Museum Pictorial No. 17. Denver Museum of Natural History, Denver.

Prairie and Grassland

Burk, W. R. 1983. Puffball usages among North American Indians. *Journal of Ethnobiology* 3: 55–62.

Jones, S. R., and R. Cushman. 2004. *North American Prairie*. Peterson Field Guide, Houghton, Mifflin, Harcourt, Boston.

Miller, O. K., Jr., R. L. Brace, and V. S. Evenson. 2005. A new subspecies of *Mycenastrum corium* from Colorado. *Mycologia* 97: 530–533.

Miller, O. K., Jr., E. Trueblood, and D. Jenkins. 1990. Three new species of *Amanita* from Southwestern Idaho and Southeastern Oregon. *Mycologia* 82: 120–128.

Nunez, M., and L. Ryvarden. 1995. *Polyporus (Basidiomycotina) and Related Genera*. Fungiflora, Oslo, Norway.

Rogerson, C. T. 1956. Kansas Mycological Notes: 1953–1954. *Transactions of the Kansas Academy of Sciences* 59(1): 39–48.

Savage, C. 2011. *Prairie, a Natural History*. Greystone Books, Vancouver, B.C., Canada.

Strickler, D. 1986. *Showy Wildflowers of the Plains, Valleys and Foothills in the Northern Rocky Mountains*. The Flower Press, Columbia Falls, Mont.

Semi-Arid Shrubland

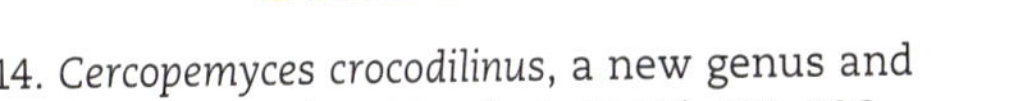

Baroni, T. J., B. R. Kropp, and V. S. Evenson. 2014. *Cercopemyces crocodilinus*, a new genus and species related to *Ripartitella*, is described from North America. *Mycologia* 106(4): 785–796.

Bates, S. 2004. Arizona Members of the Geastraceae and Lycoperdaceae (Basidiomycota, Fungi). MSc Thesis, Arizona State University, Tempe. [http://www.azfungi.org/stbates/STBThesis1.pdf, accessed May 2015]

Knight, D. H. 1996. *Mountains and Plains: The Ecology of Wyoming Landscapes*. Yale University Press, New Haven.

Kricher, J., R. T. Peterson, and G. Morrison. 1999. *A Field Guide to Rocky Mountain and Southwest Forests*. Peterson Field Guides, Houghton, Mifflin, Harcourt, Boston.

Miller, O. K., Jr., and H. H. Miller. 1988. *Gasteromycetes: Morphological and Development Features*. Mad River Press, Eureka, Calif.

Rosentreter, R., M. Bowker, and J. Belnap. 2007. *A Field Guide to Biological Soil Crusts of Western U.S. Drylands: Common Lichens and Bryophytes*. U.S. Government Printing Office, Denver, Colo.

States, J. 1990. *Mushrooms and Truffles of the Southwest*. University of Arizona Press, Tucson.

Cottonwood Riparian

Cain, K. 2007. *Meeting the Cottonwood Tree: An American Champion*. Johnson Books, Big Earth Publishing Company, Boulder, Colo.

Hesler, L. H., and A. H. Smith. 1979. *North American Species of Lactarius*. University of Michigan Press, Ann Arbor.

Kuo, M., et al. 2012. Taxonomic revision of true morels (*Morchella*) in Canada and the United States. *Mycologia* 104: 1159–1177.

Ovrebo, C., and R. Halling. 1986. *Tricholoma fulvimarginatum* (Tricholomataceae): A new species from North America associated with cottonwood. *Brittonia* 38(3): 260–263.

Smith, A. H., H. Smith, and N. Weber. 1979. *How to Know the Gilled Mushrooms*. Wm. C. Brown, Inc., Dubuque, Iowa.

Tulloss, R., and E. Moses. 1995. *Amanita populiphila*—A new species from the central United States. *Mycotaxon* 53: 455–466.

Turner, N., H. Kuhnleinh, and K. Egger. 1987. The cottonwood mushroom (*Tricholoma populinum*): A food resource of the Interior Salish Indian peoples of British Columbia. *Canadian Journal of Botany* 65: 921–927.

Vilgalys, R., A. Smith, B. L. Sun, and O. K. Miller Jr. 1993. Intersterility groups in the *Pleurotus ostreatus* complex from the continental United States and adjacent Canada. *Canadian Journal of Botany* 71: 113–128.

Ponderosa Pine Forests

Block, W. M., and D. M. Finch. 1997. *Songbird Ecology in Southwestern Ponderosa Pine Forests*. USDA FS General Technical Report RM-GTR-292, Rocky Mountain Forest and Range Station, Fort Collins, Colo.

Fiedler, C. E., and S. F. Arno. 2015. *Ponderosa: People, Fire, and the West's Most Iconic Tree*. Mountain Press Publishing Co., Missoula, Mont.

Gilbertson, R. 1974. *Fungi That Decay Pondersosa Pine*. University of Arizona Press, Tucson.

Kotter, M. M., and R. C. Farentinos. 1984. Tassel-eared squirrels as spore dispersal agents of hypogeous mycorrhizal fungi. *Journal of Mammology* 65: 684–687.

Murphy, A. 1994. *Graced by Pines: The Ponderosa Pine in the American West*. Mountain Press Publishing Co., Missoula, Mont.

Thiers, H. 1976. Boletes of the southwestern United States. *Mycotaxon* 3: 261–273.

Douglas Fir Forests

Bash, B. 1994. *Ancient Ones: The World of the Old-Growth Douglas Fir*. Gibbs Smith Publisher, Hong Kong. [children's book]

Dirks-Edmunds, J. C. 1999. *Not Just Trees: The Legacy of a Douglas-Fir Forest*. University of Washington Press, Seattle.

Maser, C., A., Claridge, J. Trappe, and C. Krebs. 2007. *Trees, Truffles, and Beasts: How Forests Function*. Rutgers University Press, Rutgers, N.J.

Maser, C., J. M. Trappe, and R. A. Nussbaum. 1978. Fungal–small mammal interrelationships with emphasis on Oregon coniferous forests. *Ecology* 59: 799–809.

Merckx, V., M. I. Bidartondo, and N. A. Hynson. 2009. Myco-heterotrophy: When fungi host plants. *Annals of Botany* 104: 1255–1261.

Mitchel, D. H., and A. H. Smith. 1976. Notes on Colorado fungi II. Species of *Armillaria* (Fr.) Kummer (Agaricales). *Mycotaxon* 4: 513–533.

Norvell, L. L., and R. L. Exeter. 2004. Ectomycorrhizal Epigeous Basidiomycetes Diversity in Oregon Coast Range *Pseudotsuga menziesii* Forests—Preliminary Observations. In: Cripps, C. L., ed. *Fungi in forest ecosystems: Diversity, ecology and systematics. Memoirs of the New York Botanical Garden* 89: 159–189.

Redhead, S. 1980. The genus *Strobilurus* (Agaricales) in Canada with notes on extralimital species. *Canadian Journal of Botany* 58: 68–83.

Thomas, D. W. 1988. The distribution of bats in different ages of Douglas-fir forests. *Journal of Wildlife Management* 52(4): 619–626.

Aspen Forests

Cripps, C. L. 1995. Mycorrhizal fungi in quaking aspen stands of Montana and Idaho. *McIlvainea* 12: 26–33.

———. 1997. The genus *Inocybe* in Montana aspen stands. *Mycologia* 89: 670–688.

———. 2001. Mycorrhizal Fungi of Aspen Forests: Natural Occurrence and Potential Applications, pp. 285–298. In: Sheppard, W., et al., eds. *Sustaining Aspen in Western Landscapes, Symposium Proceedings*, June 13–15, 2000, Grand Junction, Colo. USDA FS RMRS-P-18, Rocky Mountain Forest and Range Experiment Station, Fort Collins, Colo.

———. 2003. *Native Mycorrhizal Fungi with Aspen on Smelter-impacted Sites in the Northern Rocky Mountains: Occurrence and Potential Use in Reclamation*, pp. 193–208. National Billings Reclamation Publication, June 2003, Billings, Mont. Published by Society of Mined Land Reclamation, Lexington, Ky. [http://www.asmr.us/Publications/Conference%20Proceedings/2003/0193-Cripps.pdf, accessed May 2015]

Cripps, C. L., and O. K. Miller Jr. 1993. Ectomycorrhizal fungi associated with aspen on three sites in the north-central Rocky Mountains. *Canadian Journal of Botany* 71: 1414–1420.

Grant, M., and J. Mitton. 2010. Case Study: The glorious, golden, and gigantic quaking aspen. *Nature Education Knowledge* 3(10): 40.

Grant, M., J. Mitton, and Y. B. Linhart. 1992. Even larger organisms. *Nature* 360: 216.

Kuo, M. (2007, May). The Genus *Leccinum*. Retrieved from the *MushroomExpert.Com*. [http://www.mushroomexpert.com/leccinum.html, accessed May 2015]

Lindsey, J. P., and R. L. Gilbertson. 1978. *Basidiomycetes That Decay Aspen in North America*. J. Cramer, Germany.

Petersen, D., and B. Reynolds. 1991. *Among the Aspen*. Northland Publishing Company, Flagstaff, Ariz.

Zwinger, A. 1991. *Aspen: Blazon of the High Country*. A Peregrine Smith Book, Gibbs Smith Publisher, Layton, Utah.

———. 2002. *Beyond the Aspen Grove*, 4th edition. Johnson Publishing Company, Boulder, Colo.

Lodgepole Pine Forests

Agee, J. K. 1998. Fire and Pine Ecosystems, pp. 193–218. In: Richardson, D. M., ed. *Ecology and Biogeography of Pinus*. Cambridge Press, U.K.

Baskin, Y. 1999. Yellowstone fires: A decade later. *Bioscience* 49(2): 93–97.

Benkam, C., T. Parchman, A. Favis, and A. Siepielski. 2003. Reciprocal selection causes a coevolutionary arms race between crossbills and lodgepole pine. *The American Naturalist* 162: 182–194.

Bunyard, B. A. 2013. Matsis and wannabees: A primer on pine mushrooms. *Fungi* 6(4): 31–33.

Hosford, D., D. Pilz, R. Molina, and M. Amaranthus. 1997. *Ecology and Management of the Commercially Harvested American Matsutake Mushroom*. USDA FS General Technical Report PNW-GTR-412, Pacific Northwest Station, Portland, Ore.

Koch, P. 1996. *Lodgepole Pine in North America*. Forest Products Society, University of Michigan Press, Ann Arbor.

Nyland, R. D. 1998. Patterns of lodgepole pine regeneration following the 1988 Yellowstone fires. *Forest Ecology and Management* 111: 23–33.

Pilz, D., L. Norvell, E. Danell, and R. Molina. 2003. *Ecology and Management of Commercially Harvested Chanterelle Mushrooms*. USDA FS General Technical Report PNW-GTR-576, Pacific Northwest Research Station, Portland, Ore.

Six, D. L., J. McCutcheon, and C. Cripps. 2012. An Evolutionary Marriage. [Movie clip by C. Choate on bark beetles and their fungal symbionts; http://vimeo.com/47101945, accessed May 2015]

Burned Ground

Baker, W. 2009. *Fire Ecology in Rocky Mountain Landscapes*. Island Press, Washington, D.C.

Beug, M., A. E. Bessette, and A. R. Bessette. 2014. *Ascomycete Fungi of North America*. University of Texas Press, Austin.

Claridge, A., J. M. Trappe, and K. Hansen. 2009. Do fungi have a role as soil stabilizers and remediators after forest fires? *Forest Ecology and Management* 257: 1063–1069.

Kuo, M. 2005. *Morels*. University of Michigan Press, Ann Arbor.

———. 2008. *Morchella tomentosa*, a new species from western North America, and notes on *M. rufobrunnea*. *Mycotaxon* 105: 441–446.

———. 2012. Taxonomic revision of true morels (*Morchella*) in Canada and the United States. *Mycologia* 104: 1159–1177.

Kuo, M., D. R. Dewsbury, K. O'Donnell, M. C. Carter et al. 2011. Taxonomic revision of true morels (*Morchella*) in Canada and the United States. *Mycologia* 104: 1159–1177.

Pausas, J. G., and J. E. Keeley. 2009. A burning story: The role of fire in the history of life. *Bioscience* 59: 593–601.

Petersen, P. M. 1970. Danish fireplace fungi, an ecological investigation of fungi on burns. *Dansk Botanisk Arkiv* 27: 6–97.

Pilz, D., et al. 2007. *Ecology and Management of Morels Harvested from the Forests of Western North America*. Portland, Oregon: USDA FS General Technical Report PNW-GTR-710, Pacific Northwest Research Station, Portland, Ore.

Reinhart, K. 2008. *Yellowstone's Rebirth by Fire: Rising from the Ashes of the 1988 Wildfires*. Farcountry Press, Helena, Mont.

Richard, F., et al. 2015. True morels (*Morchella, Pezizales*) of Europe and North America: Evolutionary relationships inferred from multilocus data and a unified taxonomy. *Mycologia* 107(2): 359–382.

Rogers, M. 2004. Fire-fungi in the Columbia River Gorge, Spring 2004. *Botanical Electronic News* 329.

Romme, W. H., and D. G. Despain. 1989. Historical perspective on the Yellowstone fires of 1988. *Bioscience* 39: 696–699.

Smith, A. H., and L. R. Hesler. 1968. *The North American Species of Pholiota*. Hafner Publishing Company, New York.

Vrålstad, T. 2004. Sexual escape after fires. *Botanical Electronic News* 329.

Vrålstad, T., A. Holst-Jensen, and T. Schumacher. 1998. The postfire discomycete *Geopyxis carbonaria* (Ascomycota) is a biotrophic root associate with Norway spruce (*Picea abies*) in nature. *Molecular Ecology* 7(5): 609–616.

Weber, N. 1999. *A Morel Hunter's Companion: A Guide to True and False Morels*. Thunder Bay Press, Holt, Mich.

Wurz, T., A. Wiita, N. Weber, and D. Pilz. 2005. *Harvesting Morels after Wildfire in Alaska*. USDA FS Research Note PNW-RN-546. Pacific Northwest Research Station, Portland, Ore.

Spruce-Fir Forests

Arora, D., and L. Frank. 2014. *Boleteus rubriceps*, a new species of porcini from the southwestern USA. *North American Fungi* 9(6): 1–11.

Castellano, M., E. Cázares, B. Fondrick, and T. Dreisbach. 2003. *Handbook to Additional Fungal Species of Special Concern in the Northwest Forest Plan*. USDA FS Gen. Tech. Rep. PNW-GTR-572.

Currah, R. S., S. Hambleton, and A. Smreciu. 1988. Mycorrhizae and mycorrhizal fungi of *Calypso bulbosa*. *American Journal of Botany* 75(5): 739–752.

Dentinger, B. T., et. al. 2010. Molecular phylogenetics of porcini mushrooms (*Boletus* section *Boletus*). *Molecular Phylogenetics and Evolution* 57: 1276–1292.

Dentinger, B. T., and D. McLaughlin, 2006. Reconstructing the Clavariaceae using nuclear large subunit rDNA sequences and a new genus segregated from *Clavaria*. *Mycologia* 98: 746–762.

Drehmel, D., T. James, and R. Vilgalys. 2008. Molecular phylogeny and biodiversity of the boletes. *Fungi* 1(4): 17–23.

Michelot, D., and L. Melendez-Howell. 2003. *Amanita muscaria*: Chemistry, biology, toxicology, and ethnomycology. *Mycological Research* 107: 131–146.

Peattie, D. C. 1991 [1991]. *A Natural History of Western Trees*. Houghton Mifflin, Boston.

Tylutki, E. E. 1987. *Mushrooms of Idaho and the Pacific Northwest: Vol 2. Non-gilled Hymenomycetes*. University of Idaho Press, Moscow.

Snowbanks

Cooke, Wm. Bridge. 1955. Subalpine fungi and snowbanks. *Ecology* 36(1): 124–130.

Cripps, C. L. 2007. Snowbank Fungi of Western North America: Cold but Not Frozen. *Botanical Electronic News* 377.

———. 2009. Snowbank fungi revisited. *Fungi* 2: 47–53. [http://www.fungimag.com/spring-09-articles/13_Snow.pdf, accessed May 2015]

Halfpenny, J. C., and R. D. Ozanne. 1989. *Winter: An Ecological Handbook*. Johnson Publishing Co., Boulder, Colo.

Miller, O. K., Jr. 1965. The snowbank mushrooms in the Three Sisters Wilderness Area. *Mazama* 47(13): 38–41.

Schmidt, S. K., and D. A. Lipson. 2004. Microbial growth under the snow: implications for nutrient and alleochemical availability in temperate soils. *Plant and Soil* 259: 1–7.

Smith, A. H. 1975. *A Field Guide to Western Mushrooms*, pp. 16–17. University of Michigan Press, Ann Arbor.

High-Elevation Pines

Arno, S., and R. Hammerly. 1984. *Timberline: Mountain and Arctic Forest Frontiers*. The Mountaineers, Seattle, Wash.

Cripps, C. L. 2014. Underground Connections: Fungi and Pines in Peril. *American Forests*. [http://www.americanforests.org/our-programs/endangered-western-forests/underground-connection-fungi-and-pines-in-peril, accessed May 2015]

———. 2015. Ghost Forests. *Biosphere* 4: 32–41.

Cripps, C. L., and R. K. Antibus. 2011. Native Ectomycorrhizal Fungi of Limber and Whitebark Pine: Necessary for Forest Sustainability? pp. 37–44. In: Keane, R. E., et al., eds. Proceedings of the Hi-Five Symposium, June 28–30, 2010, Missoula, Mont. USDA FS RMRS-P-63. [http://www.fs.fed.us/rm/pubs/rmrs_p063/rmrs_p063_037_044.pdf, accessed May 2015]

Lanner, R. 1996. *Made for Each Other: A Symbiosis of Birds and Pines*. Oxford University Press, New York.

———. 2007. *The Bristlecone Book: A Natural History of the World's Oldest Trees*. Mountain Press Publishing Co., Missoula, Mont.

Lantz, G. 2010. Whitebark Pine: An Ecosystem in Peril. *American Forests Special Report*. Norman, Okla., pp. 1–8.

Mohatt, K. R., C. L. Cripps, and M. Lavin. 2008. Ectomycorrhizal fungi of whitebark pine (a tree in peril) revealed by sporocarps and molecular analysis of mycorrhizae from line forests in the Greater Yellowstone Ecosystem. *Botany* 86: 14–25.

Moser, M. M. 2004. Subalpine Conifer Forests in the Alps, the Altai, and the Rocky Mountains: A Comparison of Their Fungal Populations, pp. 151–158. In: Cripps, C. L., ed., *Fungi in forest ecosystems: Systematics, diversity and ecology*. New York Botanical Garden Press, New York.

Richardson, D. M., ed. 1998. *Ecology and Biogeography of Pinus*. Cambridge University Press, New York.

Tomback, D., S. Arno, and R. Keane. 2001. *Whitebark Pine Communities*. Island Press, Washington, D.C.

Alpine

Anderson, R. 1994. *Beartooth Country: Montana's Absaroka and Beartooth Mountains*. Magazine/American and World Geographic Publishing, Helena, Mont.

Blair, R. 1996. *The Western San Juan Mountains: Their Geology, Ecology, and Human History*. University of Colorado Press, Boulder.

Cripps, C. L., and J. Ammirati. 2010. Arctic and alpine mycology 8. *North American Fungi* 5: 1–232. [https://www.pnwfungi.org/index.php/pnwfungi/issue/view/30, accessed May 2015]

Cripps, C. L., and E. Horak. 2006. *Arrhenia auriscalpium* in arctic-alpine habitats: World distribution. Ecology, new reports from the southern Rocky Mountains, USA. *Meddelelser om Grøenland Bioscience* 56: 17–24. [*Arctic-Alpine Mycology* 6]

———. 2008. Checklist and Ecology of the Agaricales, Russulales and Boletales in the alpine zone of the Rocky Mountains (Colorado, Montana, Wyoming) at 3000–4000 m asl. *Sommerfeltia* 31: 101–123. [*Arctic-Alpine Mycology* 7]

Cripps, C. L., E. Larsson, and E. Horak. 2010. Subgenus *Mallocybe* (*Inocybe*) in the Rocky Mountain alpine zone with molecular reference to European arctic-alpine material. *North American Fungi* 5: 97–126.

Duft, J., and R. Moseley. 1989. *Alpine Wildflowers of the Rocky Mountains*. Mountain Press Publishing Co., Missoula, Mont.

Eversman, S. 1995. Lichens of alpine meadows on the Beartooth Plateau, Montana and Wyoming, U.S.A. *Arctic and Alpine Research* 27(4): 400–406.

Gellhorn, J. 2002. *Song of the Alpine: The Rocky Mountain Tundra through the Seasons*. Johnson Printing, Boulder, Colo.

Gulden, G., et al. 1985–1993. *Arctic and Alpine Mycology*, Volumes 1–4. Soppkonsulenten, Oslo, Norway.

Larsson, E., J. Vauras, and C. L. Cripps. 2014. *Inocybe leiocephala*, a species with an intercontinental distribution range-disentangling the *I. leiocephala-subbrunnea-catalaunica* morphological species complex. *Karstenia* 54: 15–39.

Miller, O. K., Jr., H. Burdsall Jr., and I. Sachs. 1980. The status of *Calvatia cretacea* in arctic and alpine tundra. *Canadian Journal of Botany* 58: 2533–2542.

Miller, O. K., Jr., and V. S. Evenson. 2001. Observations on the alpine tundra species of *Hebeloma* in Colorado. *Harvard Papers in Botany* 6: 155–162.

Moser, M., and K. McKnight. 1987. Fungi (Agaricales, Russulales) from the Alpine Zone of Yellowstone National Park and Beartooth Mountains with Emphasis on *Cortinarius*, pp. 299–317. In: Laursen, G. A., J. F. Ammirati, and S. A. Redhead, eds. *Arctic and Alpine Mycology* 2. Plenum Press, New York.

Strickler, D. 1990. *Alpine Wildflowers*. Falcon Press Publishing, Helena, Mont.

Zwinger, A., and B. Willard. 1972. *Land above the Trees: A Guide to American Alpine Tundra*. Harper and Row, New York.

Poisonous Mushrooms

Ammirati, J. F., J. A. Traquair, and P. A. Horgen. 1985. *Poisonous Mushrooms of the Northern United States and Canada*. University of Minnesota Press, Minneapolis.

Benjamin, D. R. 1995. *Mushroom Poisons and Panaceas: A Handbook for Naturalists, Mycologists, and Physicians*. W. H. Freeman and Co., New York.

Beug, M., M. Shaw, and K. Cochran. 2006. Thirty-plus years of mushroom poisoning: Summary of the approximately 2,000 reports in the NAMA case registry. *McIlvainea* 16(2): 47–68.

Lincoff, G., and D. H. Mitchel. 1977. *Toxic and Hallucinogenic Mushroom Poisoning: A Handbook for Physicians and Mushroom Hunters*. Van Nostrand Reinhold Co., New York.

Shaw, M. 1996. A closer look at mushroom poisonings. *Laboratory Medicine* 27(5): 323–327. [Rocky Mountain Poison Center, Denver, Colo.]

INDEX FOR FUNGI

Species in **bold** are treated in the book; others are mentioned. Photo pages are also in bold.

INDEX FOR PLANTS

INDEX FOR ANIMALS

Cathy Cripps

Cathy Cripps is a mycologist and associate professor in the Department of Plant Sciences and Plant Pathology at Montana State University, where she teaches and does research on fungi. She is the editor of *Fungi in Forest Ecosystems: Systematics, Diversity, and Ecology*.

Vera Evenson

Vera Evenson is curator of the Sam Mitchel Herbarium of Fungi at the Denver Botanical Gardens. She is the author of *Mushrooms of Colorado and the Southern Rocky Mountains* and past president of the Colorado Mycological Society.

Michael Kuo

Michael Kuo is a professor of English at Eastern Illinois University, the coauthor of *Mushrooms of the Midwest*, and principal developer of MushroomExpert.com.

The University of Illinois Press
is a founding member of the
Association of American University Presses.

Designed by Jennifer S. Holzner
Composed in 9/13.5 Caecilia LT Std
and 10/13 Trade Gothic LT Std
by Jennifer S. Holzner
at the University of Illinois Press
Manufactured by Bang Printing

University of Illinois Press
1325 South Oak Street
Champaign, IL 61820-6903
www.press.uillinois.edu